# Lecture Notes in Computer Science 15418

Founding Editors

Gerhard Goos
Juris Hartmanis

W0079092

The series Lecture Notes in Computer Science (LNCS), including its subseries Lecture Notes in Artificial Intelligence (LNAI) and Lecture Notes in Bioinformatics (LNBI), has established itself as a medium for the publication of new developments in computer science and information technology research, teaching, and education.

LNCS enjoys close cooperation with the computer science R & D community, the series counts many renowned academics among its volume editors and paper authors, and collaborates with prestigious societies. Its mission is to serve this international community by providing an invaluable service, mainly focused on the publication of conference and workshop proceedings and postproceedings. LNCS commenced publication in 1973.

Xiangnan He · Zhaochun Ren · Ruiming Tang
Editors

# Information Retrieval

30th China Conference, CCIR 2024
Wuhan, China, October 18–20, 2024
Revised Selected Papers

 Springer

*Editors*
Xiangnan He (ID)
University of Science and Technology
of China
Hefei, China

Zhaochun Ren (ID)
Leiden University
Leiden, The Netherlands

Ruiming Tang (ID)
Huawei Technologies Co., Ltd
Shenzhen, China

ISSN 0302-9743  ISSN 1611-3349 (electronic)
Lecture Notes in Computer Science
ISBN 978-981-96-1709-8  ISBN 978-981-96-1710-4 (eBook)
https://doi.org/10.1007/978-981-96-1710-4

This Springer imprint is published by the registered company Springer Nature Singapore Pte Ltd.
The registered company address is: 152 Beach Road, #21-01/04 Gateway East, Singapore 189721, Singapore

If disposing of this product, please recycle the paper.

# Preface

The 2024 China Conference on Information Retrieval (The 30th China Conference on Information Retrieval, CCIR 2024), organized by the Chinese Information Processing Society of China (CIPS) and hosted by Wuhan University, was successfully held in Wuhan from October 18 to 20, 2024. As the 30th installment of this conference series, CCIR 2024 has established itself as a flagship event in China's information retrieval community, witnessing the growth and innovation of information retrieval technologies.

This conference brought together renowned experts and scholars from around the world and included keynote speeches, themed workshops, poster sessions, and evaluation activities. Additionally, a young scholar forum and cutting-edge workshops on hot research topics were organized. Authors from prominent international journals and conferences (such as TOIS, SIGIR, WWW, WSDM, and CIKM) were also invited to participate in academic exchanges. To ensure the high quality of the conference, the Program Committee organized experts and scholars in the field of information retrieval to rigorously review the submitted papers. The conference received 92 submissions (including 66 Chinese-language papers and 26 English-language papers), of which 54 outstanding papers (including 43 Chinese-language papers and 11 English-language papers) were accepted, resulting in an acceptance rate of 58.7%. All accepted Chinese-language papers will be recommended for publication in relevant academic journals, while the English-language papers are published in this Springer LNCS volume.

The conference received generous support from numerous organizations and enterprises, including Tencent WeChat, Huawei Technologies, Baidu, Inc, Beijing Bayou Technology Co., and Beijing Atomic Echo Intelligent Technology Co., Ltd. We would like to extend our special thanks to these sponsors. Sincere thanks are also due to all the contributors, members of the organizing committee, conference delegates, and volunteers who contributed to the success of CCIR 2024.

The China Conference on Information Retrieval (CCIR) is the most important meeting in China in the field of information retrieval. As the flagship conference of CIPS, CCIR focuses on the development of China's internet industry and provides a broad platform for the exchange of the latest academic and technological achievements in the field of information retrieval. The theme of this year's conference, "Information Retrieval in the Era of Large Models: Challenges and Opportunities," shaped the academic activities, which included keynote speeches by renowned scholars, paper presentations, workshops on hot research topics, and a young scholar forum.

We believe this edition of the conference was a complete success.

October 2024

Xiangnan He
Zhaochun Ren
Ruiming Tang

# Organization

## Organizing Committee

## Conference Chairs

| | |
|---|---|
| Dan Wu | Wuhan University, China |
| Jiafeng Guo | Institute of Computing Techonology, CAS, China |
| Yanyan Lan | Tsinghua University, China |

## Program Committee Chairs

| | |
|---|---|
| Xiangnan He | University of Science and Technology of China, China |
| Zhaochun Ren | Leiden University, Netherlands |
| Ruiming Tang | Huawei.com, China |

## Organizing Committee Chairs

| | |
|---|---|
| Chenliang Li | Wuhan University, China |
| Fan Zhang | Wuhan University, China |

## Publicity Chair

| | |
|---|---|
| Ting Bai | Beijing University of Posts and Telecommunications, China |

## Publication Chair

| | |
|---|---|
| Xu Chen | Renmin University of China, China |

## Web Chair

Qingyao Ai                          Tsinghua University, China

## Young Scholar Forum Chairs

Weinan Zhang                        Shanghai Jiao Tong University, China
Jiawei Chen                         Zhejiang University, China

## Evaluation Chairs

Xiaofei Zhu                         Chongqing University of Technology, China
Weidong Liu                         China Mobile, China
Jingfang Xu                         Central China Normal University, China

## Sponsorship Chairs

Zhongyuan Wang                      BAAI, China
Weizhi Ma                           Tsinghua University, China

## Finance Chair

Peijie Sun                          Tsinghua University, China

# Contents

# Play to Your Strengths: Collaborative Intelligence of Conventional Recommender Models and Large Language Models

Yunjia Xi[1], Weiwen Liu[2], Jianghao Lin[1], Chuhan Wu[2], Bo Chen[2], Ruiming Tang[2], Weinan Zhang[1(✉)], and Yong Yu[1]

[1] Shanghai Jiao Tong University, Shanghai, China
{xiyunjia,chiangel,wnzhang,yyu}@sjtu.edu.cn
[2] Huawei Noah's Ark Lab, Shenzhen, China
{liuweiwen8,wuchuhan1,chenbo116,tangruimin}@huawei.com

**Abstract.** The rise of large language models (LLMs) has opened new opportunities in Recommender Systems (RSs) by enhancing user behavior modeling and content understanding. However, current approaches that integrate LLMs into RSs solely utilize either LLMs or conventional recommender models (CRMs) to generate final recommendations, without considering which data segments LLMs or CRMs excel in. To fill in this gap, we conduct experiments on MovieLens-1M and Amazon-Books datasets, and compare the performance of a representative CRM (DCNv2) and an LLM (LLaMA2-7B) on various groups of data samples. Our findings reveal that LLMs excel in data segments where CRMs exhibit lower confidence and precision, while samples where CRMs excel are relatively challenging for LLMs, requiring substantial training data and a long training time for comparable performance. This suggests potential synergies in the combination between LLMs and CRMs. Motivated by these insights, we propose Collaborative Recommendation with conventional Recommender and Large Language Model (dubbed *CoReLLa*). In this framework, we first jointly train LLMs and CRMs and address the issue of decision boundary shifts through alignment loss. Then, the resource-efficient CRMs, with a shorter inference time, handle simple and moderate samples, while LLMs process the small subset of challenging samples for CRMs. Our experimental results demonstrate that CoReLLa outperforms state-of-the-art CRMs and LLMs methods significantly, underscoring its effectiveness in recommendation tasks.

**Keywords:** Recommender System · Large Language Model

## 1 Introduction

In recent years, the emergence of large language models (LLMs) has opened up new opportunities within the realm of Recommender Systems (RSs). These LLMs, with their vast array of world knowledge and sophisticated reasoning

X. He et al. (Eds.): CCIR 2024, LNCS 15418, pp. 1–13, 2025.
https://doi.org/10.1007/978-981-96-1710-4_1

capabilities, offer a unique opportunity to revolutionize user behavior modeling and content comprehension within RSs, thereby facilitating the delivery of more accurate and personalized recommendations [2,6,19,35–38]. Current efforts have to some extent integrated large language models with recommender systems, either by injecting recommendation knowledge into LLMs [2,4,8,15,40] or by incorporating LLM knowledge into traditional recommender models (CRMs) [9,18,24,24,32,38]. However, when generating final recommendations, they adopt a binary approach, either relying entirely on LLMs or conventional recommender models. None of the previous studies have explored which specific segments of recommendation data LLMs and CRMs really excel in, neglecting potential synergies that could be leveraged to enhance recommendation quality.

In addressing this gap, we conducted experiments on two widely-used recommendation datasets, MovieLens-1M[1] and Amazon-Books[2], comparing the performance of the representative CRM (DCNv2 [33]) and LLM (LLaMA2-7B [31]) on different groups of data samples. First, we train a classical recommendation method DCNv2 on full training data and finetuned LLaMA2-7B with LoRA [12] on 10k and 100k training data following similar prompt in TALLREC [2]. Then, we adopt entropy as a confidence measurement [26], to calculate a model's prediction uncertainty. We divided the test data into three groups by ranking DCNv2's confidence in its own predictions, with the group "**1**" having the highest confidence and the group "**3**" having the lowest, as shown in Fig. 1. Finally, we compare the performance of DCNv2 (*i.e.*, **CRM**) and LLaMA2-7B finetuned on 10k and 100k training data, denoted **LLM-10k** and **LLM-100k** in Fig. 1, across these three groups. It is evident that in the first and second groups, LLM's results lag behind CRM, with only LLM-100k matching CRM's performance in the first group after trained on extensive data. This indicates that samples where CRMs excel are relatively challenging for LLMs, requiring substantial training data and long training time for comparable performance. This may be because fully trained CRMs can more easily capture certain collaborative signals unique to recommendations, such as associations between two items, while LLMs finetuned only on a subset of data may struggle to grasp such knowledge. On the contrary, in the third group, where CRMs perform the least proficiently, both LLM-10k and LLM-100k outperform CRMs. The low confidence of CRMs may come from long-tail items [3], noisy samples [14], polarizing items [3], and inconsistent user behavior [14]. On such data, LLMs can leverage its extensive world knowledge, semantic understanding, and reasoning abilities to achieve better performance, even with limited training data, like 10k.

From the above findings, a straightforward idea can be easily conceived: leveraging the strengths of each model. The resource-efficient CRMs, with a shorter inference time, can handle simple and moderate samples, while LLMs can process a small subset of challenging samples for CRMs. Here, CRMs and LLMs bear a resemblance to System 1 and System 2 in the dual-process theory [10], which elucidates cognitive processing mechanisms. In line with this framework, System 1

---

[1] https://grouplens.org/datasets/movielens/1m/.
[2] https://cseweb.ucsd.edu/~jmcauley/datasets/amazon_v2/.

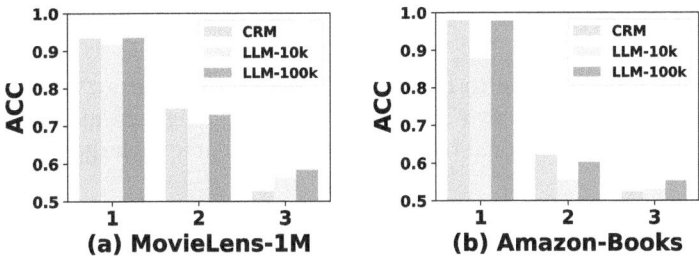

**Fig. 1.** Performance of conventional recommender models (CRMs) and Large Language Models (LLMs) on different groups.

is responsible for rapid, intuitive responses to familiar and straightforward tasks, conserving cognitive resources by swiftly executing existing routines for tasks. Conversely, System 2 engages in deliberative, analytical reasoning, activated when encountering novel or demanding situations that necessitate deeper cognitive engagement. By analogy, CRMs operate akin to System 1, efficiently managing straightforward recommendation tasks with established patterns. At the same time, LLMs function akin to System 2, employing their expansive knowledge and reasoning abilities to tackle complex recommendation challenges that may require deeper comprehension and analysis. Nonetheless, a notable issue arises when CRMs and LLMs are trained independently: their decision boundaries may diverge, that is, their boundaries between different classes or categories may be different. Merging these models without addressing this discrepancy can result in a shift in decision boundaries [34], leading to inconsistencies in how they classify or recommend items. This alteration can lead to suboptimal outcomes, undermining the efficacy of the combined approach as shown in Table 2.

Therefore, we propose Collaborative Recommendation with conventional Recommender and Large Language Model (dubbed *CoReLLa*) wherein we engage in the joint training of the two models and alignment loss for enhanced synergy. First, LLMs and CRMs are trained together with a multi-stage training strategy, due to significant differences in the parameter volumes of the two models. Additionally, a specific alignment loss is devised to mitigate the issue of decision boundary shift, thereby fostering consistency in their outputs. After training, we utilize CRM's predictions to assess the difficulty level of samples and subsequently delegate challenging samples to LLMs, ultimately amalgamating their outcomes. Our main contributions can be summarized as follows:

- We conduct the first investigation into which subset of data CRMs and LLMs excel at and find LLM performs better on data where CRMs exhibit lower confidence and CRMs can effortlessly handle samples challenging for LLMs.
- We introduce CoReLLa, where LLMs handle hard samples for CRMs and address decision boundary shift issues through multi-stage joint training and alignment loss.
- Extensive experiments demonstrate that our model outperforms SOTA CRMs and LLMs methods significantly.

## 2   Related Work

This work is closely related to LLM-enhanced recommender systems, which can be roughly classified into two categories: (1) large language models as recommenders, and (2) conventional recommenders augmented by large language models.

**Large Language Models as Recommenders.** As large language models (LLMs) demonstrate remarkable performance across various tasks in the field of natural language processing (NLP), researchers start to investigate the potential applications of LLMs to various recommendation tasks. One important line of methods is to adopt LLMs as recommenders to generate recommendations directly [2,7,21,23,36]. Due to the powerful zero-shot learning and in-context learning capabilities of LLMs, early attempts primarily focus on recommendation tasks in a zero-shot manner. For instance, ChatRec [7] employs LLMs as recommender system interfaces for conversational multi-round recommendations. Liu *et al.* [22] investigate whether ChatGPT can serve as a recommender with task-specific prompts and report the zero-shot performance. Hou *et al.* [11] further report the zero-shot ranking performance of LLMs with historical interaction data. Sanner *et al.* [28] find that LLMs provide competitive performance for pure language-based preferences in the near cold-start recommendation case in comparison to item-based CF methods. However, directly leveraging LLMs for recommendations falls behind state-of-the-art conventional recommendation algorithms, since general-purpose LLMs lack domain knowledge and collaborative signals, which are important for recommendation tasks [19]. Therefore, the focus of later work shifts to how to inject recommendation knowledge into LLMs, primarily through parameter-efficiency finetuning. For example, TALL-Rec [2] finetunes LLaMA-7B model [30] with a LoRA [12] architecture on recommendation data. ReLLa [21] design retrieval-enhanced instruction tuning by adopting semantic user behavior retrieval as a data augmentation technique and finetunes Vicuna-13B. RecRanker [23] introduces instruction-tuned LLMs for diverse ranking tasks in top-k recommendations and proposes a hybrid ranking method that ensembles various ranking tasks.

**Conventional Recommenders Augmented by Large Language Models.** Apart from directly adopting LLMs as recommenders, many researchers are also exploring the integration of open-world knowledge from LLMs into conventional recommendation models. Since large language models generally suffer from relatively long latency during inference, such an approach can enhance the recommendation effectiveness and meanwhile maintain the original inference efficiency, thereby avoiding the online inference latency issues caused by LLMs [24,37–39]. For example, KAR [38] extracts open-world knowledge from LLMs and integrates the extracted knowledge into conventional recommendation models via a hybridized expert-integrated network. LLM-Rec [24] designs various prompting strategies to elicit LLM's understanding of global and local item characteristics from GPT-3 (*text-davinci-003*), which improve the accuracy and relevance of content recommendations. Some researchers propose S&R Multi-Domain Foundation model [9], which finetunes ChatGLM2-6B [5] to extract domain invariant features for promoting performance in cold-start scenarios.

The above two types of work explore two ways of integrating LLMs and recommendations: injecting recommendation domain knowledge into LLMs and injecting LLM's knowledge into conventional recommendation models (CRM). However, regardless of which method is used, both ultimately involve using LLMs or CRMs to infer the entire dataset, without exploring whether LLMs and CRMs are better suited to certain parts of the dataset. Therefore, in this work, we explore the performance of LLMs and CRMs on different parts of the dataset and allow them to leverage their respective strengths.

## 3 Proposed Method

In this work, we focus on a core task of recommender systems, Click-Through Rate (CTR) prediction, usually formulated as a binary classification problem of predicting whether a user will click on an item. The dataset is denoted as $\mathcal{D} = \{(x_1, y_1), \ldots, (x_i, y_i), \ldots, (x_n, y_n)\}$, where $x_i$ represents the categorical features for the $i$-th instance, like item ID and user history, and $y_i$ denotes the corresponding binary label.

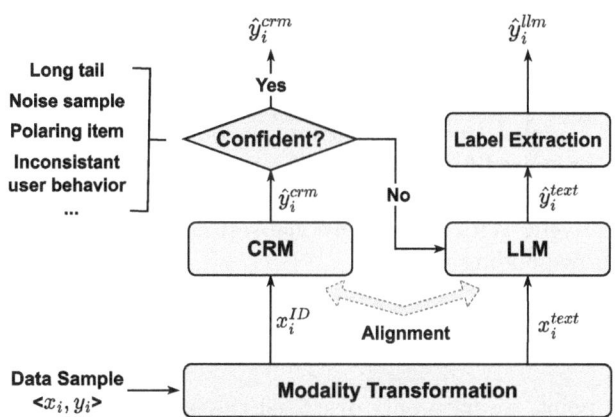

**Fig. 2.** Framework of CoReLLa.

As mentioned earlier, LLMs and CRMs each excel in different parts of recommendation data, so we have designed a framework to leverage the strengths of both CRMs and LLMs, making CRMs handle easy samples and LLMs deal with hard samples, as illustrated in Fig. 2. To mitigate decision boundary shift issues, we need to align the two models during training, for which we have designed three training stages and an alignment loss. In the inference stage, when a sample arrives, it initially leverages the CRMs branch, which has faster inference speed and lower resource consumption, to predict the result and calculate the prediction confidence. Once the confidence falls below a certain threshold, the LLM branch is activated, and the result predicted by LLMs serves as the final result. Conversely, we adopt the outcome from CRMs.

Specifically, a modality transformation module is introduced to transform the original data into recommendation and text modalities. For a data sample $<x_i, y_i>$, the recommendation modality input $x_i^{ID}$ for CRM is in a multi-field categorical data format, a one-hot vector. As for the text modality LLMs require, we utilize a hard template in Template A. Similarly, the binary label $y_i \in \{1, 0\}$ is converted into $y_i^{text} \in \{\text{"Yes"}, \text{"No"}\}$.

$x_i^{text}$ = "Below is the rating history of a user: {{user_history}}. Please predict whether the user will like {{target_item}} based on his/her rating history and the quality of the target item. You should ONLY answer no or yes. Answer: ".    (A)

Next, CRM takes $x_i^{ID}$ and generates the click probability $\hat{y}_i^{crm}$. We utilize a commonly used confidence measurement, the prediction entropy, to select hard samples, as follows:

$$s_i = -\hat{y}_i^{crm} \log \hat{y}_i^{crm} - (1 - \hat{y}_i^{crm}) \log(1 - \hat{y}_i^{crm}).    (1)$$

Typically, higher entropy indicates lower confidence in the model's predictions. The text modality $x_i^{text}$ corresponding to these hard samples is then fed into LLM. Then LLM generates the next token $\hat{y}_i^{text}$ as output, but $\hat{y}_i^{text}$ is the discrete token sampled from the distribution of LLM, not the floating-point number in $[0, 1]$ required for CTR tasks. Therefore, we extract the probabilities of "Yes" and "No" from the token distribution generated by LLMs, denoted as $a$ and $b$, respectively. With a bidimensional softmax, we can obtain $\hat{y}_i^{llm}$ which replaces the corresponding $\hat{y}_i^{crm}$

$$\hat{y}_i^{llm} = \frac{\exp(a)}{\exp(a) + \exp(b)} \in (0, 1).    (2)$$

Up to this point, we have only discussed the inference process of CoReLLa without delving into the training and optimization procedures. Next, we introduce a layer-wise alignment loss to facilitate the knowledge transformation between LLMs and CRMs, as well as aligning their outputs.

$$\mathcal{L}_{cal} = \sum_{i=1}^{n} \sum_{j=1}^{C} \|g^{llm}(h_{i,\mathcal{S}_j}^{llm}) - g^{crm}(h_{i,\mathcal{T}_j}^{crm})\|_2^{\alpha}, \ \alpha > 0.    (3)$$

In this context, both LLMs and CRMs consist of multiple layers, *i.e.*, transformer blocks in LLMs and cross net in CRMs like DCNv2. In Eq. 4, $h_{i,\mathcal{S}_j}^{llm}$ and $h_{i,\mathcal{T}_j}^{crm}$ denote the hidden state of the $i$-th sample at the $\mathcal{S}_j$-th and $\mathcal{T}_j$-th layers of LLM and CRM, respectively. Here, $\mathcal{S}$ and $\mathcal{T}$ are sets of layers chosen for LLMs and CRMs, and their size $C$ and correspondence are hyper-parameters. However, the hidden states may have different dimensions, so two transformation functions, $g^{llm}(\cdot)$ and $g^{crm}(\cdot)$, are utilized to map them into the same dimension. In practice, both $g^{llm}(\cdot)$ and $g^{crm}(\cdot)$ are a fully-connected layer. Finally, the final objective is

$$\mathcal{L} = \alpha\mathcal{L}_{llm} + \beta\mathcal{L}_{crm} + \gamma\mathcal{L}_{cal}, \qquad (4)$$

where $\mathcal{L}_{llm}$ and $\mathcal{L}_{crm}$ are the original loss of LLMs and CRMs, $\alpha \in [0, \infty)$, $\beta \in [0, \infty)$, and $\gamma \in [0, \infty)$ denotes the hyper-parameters that control the weight of losses. To enhance the mix-up strategy of LLMs and CRMs, we employ a multi-stage training approach:

- **Stage 1 (CRM warm-up training)**: In this phase, we train CRMs with the entire training set as an initialization. To achieve satisfactory results, CRMs often require substantial training data, a quantity challenging to attain during joint training with LLMs. Thus, in this stage, $\alpha = \gamma = 0$ and $\beta = 1$.
- **Stage 2 (Joint training with alignment)**: During this stage, we randomly select a small subset of training data, such as 1%, to simultaneously train LLMs and CRMs while calibrating their results. Here $\alpha$, $\beta$, and $\gamma$ are non-zero. In experiments, $\alpha = \beta = 1$ and $\gamma = 0.1$.
- **Stage 3 (LLM continue training)**: The previous stage has observed a seesaw phenomenon in the optimization of CRMs and LLMs—as LLMs continue to improve, CRMs experience a decline in performance. Therefore, after achieving favorable results in CRMs during joint training, we cease joint training and proceed to continue training LLMs with another randomly sampled subset from training data. Here, $\alpha = 1$ and $\beta = \gamma = 0$.

## 4 Experiments

### 4.1 Setup

Our experiments are conducted on two public datasets, MovieLens-1M[3] and Amazon-Book[4]. **MovieLens-1M** comprises 1 million ratings provided by 6000 users for 4000 movies. We follow common practices [27,41] in CTR prediction for data processing. The ratings are transformed into binary labels, with ratings of 4 and 5 labeled as positive, and the rest as negative. The data samples are sorted by their global timestamps, with the first 80% selected as the training set, the middle 10% as the validation set, and the final 10% as the test set. The models receive inputs consisting of item ID, user ID, and associated attribute features of users and items. **Amazon-Book** [25] is derived from the "Books" category of the Amazon Review Dataset, and it undergoes filtration to exclude less-interacted users and items. The ratings of 5 are considered positive, while the rest are deemed negative. The preprocessing of Amazon-Book is akin to those applied to MovieLens-1M, with the difference being the absence of user features.

---

[3] https://grouplens.org/datasets/movielens/1m/.
[4] https://cseweb.ucsd.edu/~jmcauley/datasets/amazon_v2/.

Click-Through Rate (CTR) prediction aims to predict the likelihood of a user clicking on an item, which is a core task in recommendation systems. Therefore, our experiments are conducted based on CTR prediction tasks. We select several representative traditional CTR prediction models such as DCNv2 [33], FiBiNet [13], AutoInt [29], xDeepFM [17], Fi-GNN [16], *etc..*, as baselines. For instance, **xDeepFM** [17] leverages the power of both deep network and Compressed Interaction Network to generate feature interactions at the vector-wise level. **DCNv2** [33] is an improved framework of DCN which is more practical in large-scale industrial settings. **FiBiNet** [13] can dynamically learn the feature importance by Squeeze-Excitation network and fine-grained feature interactions by bilinear function. **FiGNN** [16] converts feature interactions into modeling node interactions on the graph for modeling feature interactions in an explicit way. **AutoInt** [29] adopts a self-attentive neural network with residual connections to model the feature interactions explicitly. Additionally, we also compare recommendation models based on LLM, including P5 [8], TALLREC [2], and CTRL [15], and adapted them to CTR prediction tasks. For example, **P5** [8] is a text-to-text paradigm that unifies recommendation tasks and learns different tasks with the same language modeling objective during pretraining. **TALL-Rec** [2] finetunes LLaMa-7B [30] with a LoRA architecture on recommendation tasks and enhances the recommendation capabilities of LLMs in few-shot scenarios. In our experiment, we implement TALLRec with LLaMa-2-7B-chat[5], since it has better performance and ability of instruction following. We employ widely-used $ACC$ (Accuracy), $AUC$ (Area under the ROC curve) and $LogLoss$ (binary cross-entropy loss) as evaluation metrics following [29,33,41]. A higher AUC value or a lower Logloss value, even by a small margin (*e.g.*, 0.001), can be viewed as a significant improvement in CTR prediction performance, as indicated by previous studies [17,20,33].

As for our model, we opt for DCNv2 [33] as CRMs and LLaMa2-7b-chat as LLMs. Firstly, DCN undergoes warm-up training on the entire dataset. Subsequently, DCN and LLaMa2-7b (finetuned with LoRA) are jointly trained on 20–30k data samples which are randomly selected from the training set. Finally, LLaMa2-7b is trained independently on the other 20–30k randomly selected data samples. Other parameters, such as batch size, learning rate, and weight decay are determined through grid search to achieve the best results. For fair comparisons, the parameters of the backbone model and the baselines are also tuned to achieve their optimal performance.

### 4.2   Overall Performance

We evaluate our proposed models and baseline models with AUC (Area under the ROC curve), ACC (Accuracy) and LogLoss (binary cross-entropy loss) in Table 1. Based on the experimental results, the following conclusions can be drawn: (1) our model CoReLLa significantly outperforms CRMs and PLM-based models. For instance, on Amazon-Books, our proposed CoReLLa demonstrates a

---

[5] https://huggingface.co/meta-llama/Llama-2-7b-chat-hf.

notable improvement over the best baselines, with a 1.38% reduction in Logloss and a 1.03% increase in ACC. On the MovieLens-1M dataset, CoReLLa also demonstrates an improvement of 0.72% in AUC and 1.08% in ACC. This indicates that CoReLLa successfully integrated the strengths of LLMs and CRMs, yielding superior results than both types of models. (2) Pure PLM-based recommendations such as P5 and TALLREC often fall short compared to CRMs like FiBiNet and AutoInt. This also validates our conclusion in Fig. 1 that LLMs do not surpass CRMs in most samples. This indicates that it remains challenging for LLMs to surpass well-designed CRMs, and further integration with domain knowledge in the recommendation field may be required. However, utilizing larger language models tends to be more effective than smaller ones. For example, TALLREC based on LLaMa-7B outperforms CTRL based on BERT, suggesting that recommendation can benefit from larger language models.

**Table 1.** Overall performance on two benchmark datasets. We underline the second-best value and denote the best result in bold, whose improvements are statistically significant with $p < 0.05$ against best baselines denoted by *.

| Model | MovieLens-1M | | | Amazon-Books | | |
|---|---|---|---|---|---|---|
| | AUC | Logloss | ACC | AUC | Logloss | ACC |
| DCNv2 | 0.7939 | 0.5469 | <u>0.7230</u> | 0.8255 | 0.5012 | 0.7481 |
| xDeepFM | 0.7925 | 0.5449 | 0.7210 | 0.8253 | 0.5021 | 0.7481 |
| FiBiNet | <u>0.7947</u> | 0.5442 | 0.7228 | 0.8254 | 0.5018 | 0.7479 |
| AutoInt | 0.7909 | 0.5472 | 0.7214 | <u>0.8256</u> | <u>0.5010</u> | 0.7480 |
| DeepFM | 0.7940 | <u>0.5439</u> | 0.7225 | 0.8252 | 0.5015 | <u>0.7483</u> |
| FiGNN | 0.7921 | 0.5464 | 0.7209 | 0.8224 | 0.5046 | 0.7458 |
| P5 | 0.7902 | 0.5516 | 0.7174 | 0.7986 | 0.5320 | 0.7275 |
| TALLREC | 0.7931 | 0.5463 | 0.7209 | 0.8239 | 0.5060 | 0.7436 |
| CTRL | 0.7929 | 0.5465 | 0.7218 | 0.7996 | 0.5297 | 0.7253 |
| **CoReLLa** | **0.8001\*** | **0.5402\*** | **0.7308\*** | **0.8303\*** | **0.4941\*** | **0.7558\*** |

## 4.3   Ablation Study

In this section, we explore how different training stages and the mix-up strategy of CoReLLa impact the final results. We design four variants and conduct experiments on two datasets, with the results presented in Tables 1 and 2. In the table, "**w/o S1**", "**w/o S2**", and "**w/o S3**" respectively denote the exclusion of training stages 1 (CRM warm-up training), stage 2 (joint training with alignment), and stage 3 (LLM continue training). "**w/o mix**" indicates generating recommendations by CRMs after joint training in stage 2 without mix-up strategy.

**Table 2.** Performance of different variants on two datasets.

| Variants | MovieLens-1M | | | Amazom-Books | | |
|---|---|---|---|---|---|---|
| | AUC | Logloss | ACC | AUC | Logloss | ACC |
| w/o S1 | 0.6511 | 0.6611 | 0.6029 | 0.8046 | 0.5303 | 0.7354 |
| w/o S2 | 0.7941 | 0.5467 | 0.7259 | 0.8265 | 0.4988 | 0.7486 |
| w/o S3 | 0.7990 | 0.5400 | 0.7284 | 0.8277 | 0.4966 | 0.7468 |
| w/o mix | 0.7982 | 0.5410 | 0.7276 | 0.8285 | 0.4959 | 0.7501 |
| CoReLLa | **0.8001** | **0.5402** | **0.7308** | **0.8303** | **0.4941** | **0.7558** |

From the table, we observed the most significant decrease in model performance when excluding S1, primarily due to the pivotal role of CRMs in our framework. The confidence level of CRMs is used to determine which samples require processing by LLMs, and CRMs also handle the majority of the data. Hence, the entire framework relies on high-quality CRMs. Typically, CRMs trained on the full dataset achieve better results, and removing Stage 1 leads to poorer performance since CRMs are only trained on a small amount of data, consequently resulting in a decrease in overall model performance. The removal of S2 also results in a notable decline, even inferior to the performance of baseline CRMs, especially on AUC. This indicates that without joint training and alignment, the simple combination of LLMs and CRMs trained separately may experience decision boundary shifts and reduced effectiveness. While the exclusion of the mix-up strategy leads to a certain decline in performance, it still outperforms the baseline CRMs. This implies that during joint training, LLMs impart knowledge to CRMs, enhancing CRMs' performance.

## 5 Conclusion

In this paper, we point out that current works solely use either LLMs or CRMs for recommendations, overlooking their distinct strengths. Therefore, we conduct the first experiments to compare the performance of CRMs and LLMs on various data segments. Findings show LLMs excel where CRMs exhibit lower confidence, suggesting synergies in their combination. Thus, we propose CoReLLa, which jointly trains LLMs and CRMs via a multi-stage training strategy and alignment loss to address the issues of decision boundary shifts. CoReLLa outperforms state-of-the-art CRMs and LLMs methods, highlighting its effectiveness in recommendations.

**Acknowledgment.** The Shanghai Jiao Tong University team is partially supported by National Natural Science Foundation of China (62177033, 62076161) and Shanghai Municipal Science and Technology Major Project (2021SHZDZX0102). The work is also sponsored by Huawei Innovation Research Program. We thank MindSpore [1] for its partial support. The author Yunjia Xi is also supported by Wu Wen Jun Honorary Doctoral Scholarship.

**Disclosure of Interests.** The authors have no competing interests to declare that are relevant to the content of this article.

# References

1. Mindspore (2020). https://www.mindspore.cn/
2. Bao, K., Zhang, J., Zhang, Y., Wang, W., Feng, F., He, X.: TALLREC: an effective and efficient tuning framework to align large language model with recommendation. arXiv preprint arXiv:2305.00447 (2023)
3. Cleger-Tamayo, S., Fernández-Luna, J.M., Huete, J.F., Tintarev, N.: Being confident about the quality of the predictions in recommender systems. In: Serdyukov, P., et al. (eds.) ECIR 2013. LNCS, vol. 7814, pp. 411–422. Springer, Heidelberg (2013). https://doi.org/10.1007/978-3-642-36973-5_35
4. Cui, Z., Ma, J., Zhou, C., Zhou, J., Yang, H.: M6-REC: generative pretrained language models are open-ended recommender systems. arXiv preprint arXiv:2205.08084 (2022)
5. Du, Z., et al.: GLM: general language model pretraining with autoregressive blank infilling. In: Proceedings of the 60th Annual Meeting of the Association for Computational Linguistics (Volume 1: Long Papers), pp. 320–335 (2022)
6. Friedman, L., et al.: Leveraging large language models in conversational recommender systems. arXiv preprint arXiv:2305.07961 (2023)
7. Gao, Y., Sheng, T., Xiang, Y., Xiong, Y., Wang, H., Zhang, J.: Chat-REC: towards interactive and explainable LLMs-augmented recommender system. arXiv preprint arXiv:2303.14524 (2023)
8. Geng, S., Liu, S., Fu, Z., Ge, Y., Zhang, Y.: Recommendation as language processing (RLP): a unified pretrain, personalized prompt and predict paradigm (p5). In: RecSys, pp. 299–315 (2022)
9. Gong, Y., Ding, X., Su, Y., Shen, K., Liu, Z., Zhang, G.: An unified search and recommendation foundation model for cold-start scenario. In: Proceedings of the 32nd ACM International Conference on Information and Knowledge Management. CIKM '23, pp. 4595–4601 (2023)
10. Groves, P.M., Thompson, R.F.: Habituation: a dual-process theory. Psychol. Rev. **77**(5), 419 (1970)
11. Hou, Y., et al.: Large language models are zero-shot rankers for recommender systems. arXiv preprint arXiv:2305.08845 (2023)
12. Hu, E.J., et al.: Lora: low-rank adaptation of large language models. arXiv preprint arXiv:2106.09685 (2021)
13. Huang, T., Zhang, Z., Zhang, J.: Fibinet: combining feature importance and bilinear feature interaction for click-through rate prediction. In: RecSys, pp. 169–177 (2019)
14. Joorabloo, N., Jalili, M., Ren, Y.: Improved recommender systems by denoising ratings in highly sparse datasets through individual rating confidence. Inf. Sci. **601**, 242–254 (2022)
15. Li, X., Chen, B., Hou, L., Tang, R.: CTRL: connect tabular and language model for CTR prediction. arXiv preprint arXiv:2306.02841 (2023)
16. Li, Z., Cui, Z., Wu, S., Zhang, X., Wang, L.: Fi-GNN: modeling feature interactions via graph neural networks for CTR prediction. In: CIKM, pp. 539–548 (2019)
17. Lian, J., Zhou, X., Zhang, F., Chen, Z., Xie, X., Sun, G.: XdeepFM: combining explicit and implicit feature interactions for recommender systems. In: KDD, pp. 1754–1763 (2018)

18. Lin, J., et al.: Clickprompt: CTR models are strong prompt generators for adapting language models to CTR prediction. arXiv preprint arXiv:2310.09234 (2023)
19. Lin, J., et al.: How can recommender systems benefit from large language models: a survey. arXiv preprint arXiv:2306.05817 (2023)
20. Lin, J., et al.: Map: a modelagnostic pretraining framework for click-through rate prediction. In: Proceedings of the 29th ACM SIGKDD Conference on Knowledge Discovery and Data Mining, pp. 1384–1395 (2023)
21. Lin, J., et al.: ReLLa: retrieval-enhanced large language models for lifelong sequential behavior comprehension in recommendation. arXiv preprint arXiv:2308.11131 (2023)
22. Liu, J., Liu, C., Lv, R., Zhou, K., Zhang, Y.: Is ChatGPT a good recommender? A preliminary study. arXiv preprint arXiv:2304.10149 (2023)
23. Luo, S., et al.: Recranker: instruction tuning large language model as ranker for top-k recommendation. arXiv preprint arXiv:2312.16018 (2023)
24. Lyu, H., Jiang, S., Zeng, H., Xia, Y., Luo, J.: LLM-REC: personalized recommendation via prompting large language models. arXiv preprint arXiv:2307.15780 (2023)
25. Ni, J., Li, J., McAuley, J.: Justifying recommendations using distantly-labeled reviews and fine-grained aspects. In: Proceedings of the 2019 Conference on Empirical Methods in Natural Language Processing and the 9th International Joint Conference on Natural Language Processing (EMNLP-IJCNLP), pp. 188-197 (2019)
26. Park, L.A.F., Simoff, S.: Using entropy as a measure of acceptance for multi-label classification. In: XIV (2015)
27. Qin, J., Zhang, W., Wu, X., Jin, J., Fang, Y., Yu, Y.: User behavior retrieval for click-through rate prediction. In: Proceedings of SIGIR, pp. 2347-2356 (2020)
28. Sanner, S., Balog, K., Radlinski, F.,Wedin, B., Dixon, L.: Large language models are competitive near cold-start recommenders for language-and item-based preferences. In: Proceedings of the 17th ACM Conference on Recommender Systems, pp. 890–896 (2023)
29. Song, W., et al.: Autoint: automatic feature interaction learning via self-attentive neural networks. In: CIKM, pp. 1161–1170 (2019)
30. Touvron, H., et al.: Llama: open and efficient foundation language models. arXiv preprint arXiv:2302.13971 (2023)
31. Touvron, H., et al.: Llama 2: open foundation and fine-tuned chat models. arXiv preprint arXiv:2307.09288 (2023)
32. Wang, H., et al.: Flip: towards fine-grained alignment between id-based models and pretrained language models for CTR prediction. arXiv e-prints pp. arXiv-2310 (2023)
33. Wang, R., et al.: DCN V2: improved deep & cross network and practical lessons for web-scale learning to rank systems. In: WWW, pp. 1785–1797 (2021)
34. Wang, Z., et al.: Augmentation with projection: towards an effective and efficient data augmentation paradigm for distillation. In: ICML (2023)
35. Wen, M., et al.: Large sequence models for sequential decision-making: a survey. Front. Comput. Sci. **17**(6), 176349 (2023)
36. Xi, Y., et al.: MemoCRS: memoryenhanced sequential conversational recommender systems with large language models. arXiv preprint arXiv:2407.04960 (2024)
37. Xi, Y., et al.: Efficient and deployable knowledge infusion for open-world recommendations via large language models. arXiv preprint arXiv:2408.10520 (2024)
38. Xi, Y., et al.: Towards open-world recommendation with knowledge augmentation from large language models. arXiv preprint arXiv:2306.10933 (2023)

39. Xi, Y., et al.: A decoding acceleration framework for industrial deployable LLM-based recommender systems. arXiv preprint arXiv:2408.05676 (2024)
40. Zhang, Y., et al.: Language models as recommender systems: evaluations and limitations. In: I (Still) Can't Believe It's Not Better! NeurIPS 2021 Workshop (2021)
41. Zhou, G., et al.: Deep interest network for click-through rate prediction. In: Proceedings of the 24th ACM SIGKDD International Conference on Knowledge Discovery and Data Mining, pp. 1059–1068 (2018)

# A Dual-Aligned Model for Multimodal Recommendation

Kangning Zhang[1], Yingjie Qin[2], Ruilong Su[2], Yifan Liu[1],
Weinan Zhang[1]([✉]), and Yong Yu[1]

[1] Shanghai Jiaotong University, Shanghai, China
{zhangkangning,sjtulyf123,wnzhang,yyu}@sjtu.edu.cn
[2] XiaoHongShu, Shanghai, China
{huanling,daisuke}@xiaohongshu.com

**Abstract.** This paper focuses on the effective utilization of multimodal information in multimodal recommendation systems. Previous research in this field can be broadly categorized into two approaches: explicit and implicit utilization of multimodal information. Explicit utilization directly incorporates modality features extracted from items into the network computation. However, this approach may suffer from suboptimal performance when the available multimodal information is limited. On the other hand, implicit utilization learns the representations of users and items by constructing auxiliary structures, such as multimodal graphs. These structures help implicitly recall relevant items from semantic perspective. However, this approach belongs to the local utilization of multimodal features due to the Top-$k$ sampling and is not well suited for handling large amounts of modal data. To address these limitations, we propose a novel model called **DAMORE** (**D**ual-**A**ligned Model for **M**ultim**O**dal **RE**commendation), which combines explicit and implicit utilization to help model achieve SOTA performance in scenarios with sparse multimodal features and large data scales. In implicit alignment, to provide a boarder perspective for local utilization, we introduce a novel user-user graph and combine it with item-item graph from previous works. In explicit alignment, we employ a self-supervised learning method to align item representations with multimodal features. Furthermore, previous approaches allow multimodal features to be updated during training, leading to semantic shift of original modal information. Based on this phenomenon, we introduce Modal Feature Persistence (MFP) to ensure feature stability during training. We conduct extensive experiments on three public datasets, and our model achieves state-of-the-art (SOTA) results. The code is accessible at this url.

**Keywords:** Information systems · Multimodal recommendation

## 1 Introduction

With the high prevalence of the Internet, multimedia recommendation has been a core service to help users identify items of interest on multimedia content

X. He et al. (Eds.): CCIR 2024, LNCS 15418, pp. 14–27, 2025.
https://doi.org/10.1007/978-981-96-1710-4_2

sharing platforms. Nevertheless, the vast amount of multimodal content information associated with items remains largely unexplored. The main challenge in multimodal recommendation systems is how to effectively utilize multimodal information and integrate it into recommendation tasks. To improve the recommendation accuracy, recent works on multimodal recommendation have studied effective means to integrate items' multimodal information into the traditional user-item recommendation paradigm. We claim that previous methods can be broadly categorized into two types: those that explicitly utilize multimodal information and those that implicitly utilize it.

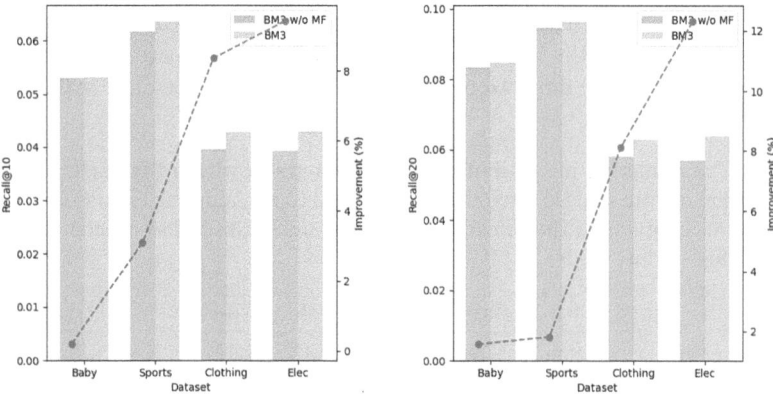

**Fig. 1.** Comparison between BM3 w/o MF and BM3. "BM3 w/o MF" means removing all modal-related features and parameters of BM3. The red dashed line indicates the performance improvement ratio of BM3 relative to BM3 w/o MF. As multimodal data becomes more abundant, BM3 achieves greater performance boost through explicit utilization of multimodal features. (Color figure online)

The "explicitly utilize" refers to using multimodal features as explicit supervisory signals for model learning. VBPR [6] explicitly concatenates modal features with item embeddings. MMGCN [16], DualGNN [16] use the power of graph neural networks [16,19] for capturing high-order semantics information by using modal features as initialed embedding. BM3 [24] incorporates self-supervised learning paradigm [1,3,4] to align item representation and multimodal features. Experimentally, we find that explicit utilization performs well only with sufficient multimodal data. In other words, when multimodal information is limited, it is difficult to provide sufficient learning signals for recommendation models, which results in suboptimal performance. In Fig. 1, we demonstrated that as the modal feature scale increases, BM3 [24] achieves increasingly performance improvements through explicitly utilizing multimodal information. In addition, previous works have a serious problem of "**multimodal information shift**". That is, multimodal features is updated constantly during the training process and therefore gradually deviates from the initial value.

In contrast, "implicitly utilize" refers to using the multimodal features to construct auxiliary structure (i.e. multimodal graphs) instead of incorporating them into the computation of the neural network. Representative works in this category include LATTICE [20] and FREEDOM [23] which achieve the implicit utilization by constructing item-item graphs using multimodal information. When interaction information is limited, implicit modal graphs can effectively provide additional semantic information and achieve better performance than explicit utilization. However, due to retaining only the top-$k$ most similar items in item-item graph, implicit utilization is limited to local information of multimodal features. As data scale increases, the performance of the model is affected. We directly use the item-item graph to recall the top-k most similar items and evaluate the recall rate on the test set. The results in Table 1 demonstrate that as dataset size increases, the item-item recall rate gradually decreases, further indicating the performance limitations of item-item graph as the data scale expands.

**Table 1.** The reduction in recall of the item-item graph within FREEDOM [23] with the expansion of data scale is attributed to the local top-$k$ sampling employed in implicit utilization. Consequently, the efficacy of implicit models diminishes concomitantly with the escalation of data scale.

|  | Baby | Sports | Clothing | Elec |
|---|---|---|---|---|
| $k = 10$ | 12.70% | 11.31% | 10.50% | 4.40% |
| $k = 20$ | 15.24% | 14.86% | 12.28% | 6.40% |

Therefore, relying solely on either explicit or implicit utilization of multimodal information may only achieve suboptimal results. On the basis of these ideas, we propose DAMORE to successfully combine explicit and implicit utilization of multimodal features. When multimodal features are limited, explicit utilization may not provide sufficient learning signals, in such cases, constructing modal graphs implicitly help model learning. On the other hand, as the dataset scale expands, relying solely on implicit sampling of top-k items may not capture the global picture accurately. Explicit utilization, however, can assist DAMORE in such scenario. The combination enables DAMORE to achieve better performance compared with previous works in scenarios with sparse multimodal features and larger dataset scales.

In the *implicit alignment*, in order to expand the receptive field for multimodal features, we construct modal item-item graph and a novel modal **user-user graph**. Our experimental analysis of implicit semantic recall on datasets of different scales verifies our hypothesis that there is a strong correlation between the recall rate of modal graphs and the model's performance. Further, the user-user graph is more helpful than item-item graph in our experiments. We think that aggregating multimodal features alleviates the noise from original modal features and the constructed user-user graph is more robust than item-item graph. In the *explicit alignment*, we adopt a self-supervised learning method to align modal representations and id embeddings. Specially, we use a Multi-View Generator to generate dual views of latent vectors and leverage contrastive learning for alignment. In previous works [23,24], the multimodal features are set as learnable embeddings so that the multimodal information can be fine-tuned in training to better fit in with the recommendation task. However, such method may gradually lose original multimodal information and cause the problem of multimodal information shift. To make a trade-off between multimodal and behavioral information, we propose the concept of Modality Feature Persistence (MFP) to ensure that multimodal features can be updated but remain relatively stable throughout the training process. In our method, the multimodal features are mildly updated at first and then quickly turn to convergence, which means that the original multimodal information remain unlost and the moduled MFP can be integrated into other methods easily.

We summarize our main contributions as follows:

- We revealed the limitations of the latest explicit and implicit utilization models through experiments and subsequently proposed DAMORE, a dual align-ment approach that effectively integrates both implicit and explicit utilization of multimodal features in Multimodal Recommendation.
- In the implicit alignment, we additionally introduce modal user-user graph for broader receptive field in multimodal features and surprisingly discover user-user graph provides greater assistance to the model's performance compared to the item-item graph.
- We introduce the concept of Modal Feature Persistence (MFP) to address the issue of modal feature shift, which maintains multi-modal information effectively and achieves better performance.

Finally, We conduct extensive experiments on three public datasets and our method achieves state-of-the-art (SOTA) results.

## 2  Related Works

### 2.1  Implicit Utilization Models

Implicit utilization models use the multimodal features to construct auxiliary structures (i.e. multimodal graphs) instead of directly incorporating them into the computation. GRCN [18] scores the affinity between users and items through implicit utilization of modal features. LATTICE [20] introduces an item-item graph based on modality information. FREEDOM [23] uses the same item-item graph as LATTICE but freezes it during training and introduces the degree-sensitive edge pruning techniques to denoise the user-item graph. Although these methods implicitly introduce modal information and improve performance in some scenarios, this benefit is weakened when interaction data increases. In our work, we implicitly construct a modal item-item graph and modal user-user graph to increase the local utilization of modal features. DREAM [21] extends the modal relation graphs to user-user perspective.

## 2.2   Explicit Utilization Models

Explicit utilization models directly incorporating the multimodal features extracted from items through Vision Networks [2,5,13] or Language Models [10,11,15] into the computation. VBPR [6,17] represents an item by fusing the latent visual features. Based on VBPR, Deepstyle [8] augments the representations of items with both visual and style features. MMGCN [19] constructs a bipartite user-item modality-specific graph to capture neighbor information. DualGNN [16] is based on MMGCN and introduces a user concurrence graph to capture user preferences for modal features. Although these methods implicitly introduce modal information and improve performance in some scenarios, this benefit is weakened when interaction data increases. In our work, we implicitly utilize the multimodal feature to construct a modal item-item graph and modal user-user graph to increase the local utilization of modal features. Another line of works [9,14,24] successfully incorporate various alignment tasks into explicit utilization in multimodal recommendation. SLMRec [14] employs self-supervised learning in a graph neural network to learn the underlying relationships between interactions. BM3 [24] proposes a novel self-supervised learning approach that addresses the issues of high computational cost and incorrect supervision signals. AlignRec [9] breaks down the recommendation task into three alignment objective, with each governed by unique objective functions. In this work, we adopt an unsupervised learning method inspired by BM3 [24] and at the same time address the problem of multimodal information shift present in prior works.

## 3   DAMORE

In this section, we introduce the details of our proposed DAMORE model, which encompasses dual alignments, as illustrated in Fig. 2.

### 3.1   Preliminary

Given a set of M users $u \in \mathcal{U}$, a set of N items $i \in \mathcal{I}$, we denote the historical behavior data as $R \in \mathbb{R}^{M \times N}$, where $R_{ui} = 1$ if user $u$ interacted with item $i$, otherwise $R_{ui} = 0$. Also, We model the dynamic relations of user interactions as a user-item bipartite graph $\mathcal{G} = \{\mathcal{U}, \mathcal{I}, \mathcal{E}\}$, where we regard the historical interactions as the set of edges in the graph denoted by $\mathcal{E} = \{(u, i) | u \in \mathcal{U}, i \in \mathcal{I}, R_{ui} = 1)\}$. Each item is associated with multimodal content information $m \in \{v, t\}$, where $v$ and $t$ represent visual and textual features respectively. We denote the modality feature for an item $i$ as $x_i^m \in \mathbb{R}^{d_m}$, where $d_m$ denotes the feature dimension of modality $m$. The aim of multimodal recommendation is to accurately predict users' preferences by ranking items for each user according to predicted scores $y_{ui}$.

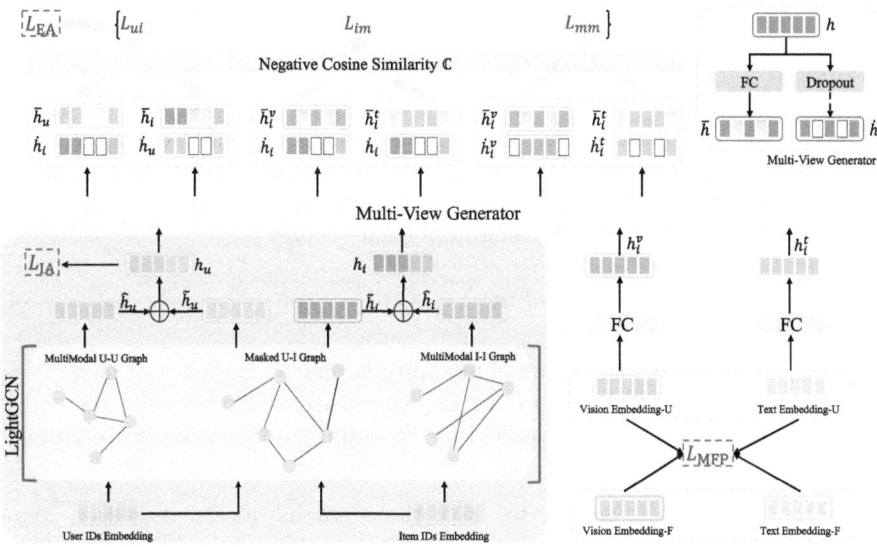

**Fig. 2.** The structure overview of the proposed DAMORE can be viewed as implicit alignment(IA), masked as $L_{IA}$ and explicit alignment(EA), masked as $L_{EA}$. We also try to maintain modal feature consistency by $L_{MFP}$, in which X-F represents a frozen multimodal vector and X-U indicates a learnable embedding.

## 3.2   Implicit Alignment

**Modal Graphs and Behavior Graph Construction.** For every item, there exists a multimodal feature vector $x_i^m \in R^{d_m}$ in each specific modality. Let $I_u = \{i | (u, i) \in \mathcal{E}\}$ denote the set of items which a user $u$ has interacted with. We obtain the user modal feature in a specific modality through mean pooling:

$$y_u^m = \frac{1}{|I_u|} \sum_{i \in I_u} x_i^m \tag{1}$$

We use the cosine similarity function to calculate the similarity score $S_{ij}^m$ between each pair of items $i$ and $j$ in modality $m$:

$$S_{ij}^m = \frac{(x_i^m)^T x_j^m}{||x_i^m|| ||x_j^m||}, \quad J_{ij}^m = \frac{(y_i^m)^T y_j^m}{||y_i^m|| ||y_j^m||} \tag{2}$$

The similarity between item $i$ and item $j$ in a specific modality is measured by the matrix $S^m \in \mathbb{R}^{N \times N}$, while the similarity between user $i$ and user $j$ is measured by the matrix $J^m \in \mathbb{R}^{M \times M}$. Although $S^m$ and $J^m$ have different dimensions, they undergo the same computation, we take $J^m$ as example.

We only keep the top-$k$ similar users of every user and convert the weighted graph into an unweighted graph:

$$\hat{J}_{ij}^m = \begin{cases} 1, & J_{ij}^m \in \text{topk}(J_i^m) \\ 0, & \text{otherwise} \end{cases} \tag{3}$$

Next, we normalize the discretized adjacency matrix as $\tilde{J}^m = (D^m)^{-\frac{1}{2}}\hat{J}^m(D^m)^{-\frac{1}{2}}$, where $D^m \in \mathbb{R}^{M \times M}$ is the diagonal degree matrix of $\hat{J}^m$ and $D_{ii}^m = \sum_j \hat{J}_{ij}^m$. Then, we construct the modal user-user graph by aggregating the graphs from each modality using importance scores $\alpha_m$:

$$J = \sum_{m \in \mathcal{M}} \alpha_m \tilde{J}^m \tag{4}$$

where $\alpha_m$ is the importance score of modality $m$ and $\mathcal{M}$ is the set of modalities.

Following FREEDOM [23], we use degree-sensitive edge pruning to create a masked user-item interaction graph $\hat{A}$ for denoising user-item bipartite graph.

**Implicit Alignment.** So far, we have three graphs: modal user-user, item-item graph and behavior user-item graph. We aim to learn dual representations of users and items, hoping to capture information from modal and behavioral aspects. Specifically, we perform graph convolutions on graphs $S$, $J$, and $\hat{A}_m$.

The graph convolution over the item-item graph and user-user graph is:

$$\tilde{h}_i^l = \sum_{j \in \mathcal{N}(i)} S_{ij}\tilde{h}_j^{l-1}, \quad \tilde{h}_u^l = \sum_{j \in \mathcal{N}(u)} J_{uj}\tilde{h}_j^{l-1} \tag{5}$$

Here $\mathcal{N}(i)$ represents the set of neighbors of item $i$, $\tilde{h}_i^l \in \mathbb{R}^d$ represents the $l$-th layer representation of item $i$, $\tilde{h}_i^0$ represents item $i$ embedding. We stack $L_{ii}$ convolutional layers on the item-item graph $S$ and obtain the last layer representation as the representation $\tilde{h}_i \in \mathbb{R}^d$ of item $i$.

In the user-item graph, we perform $L_{ui}$ convolutional operations on $\hat{A}$. Specifically, we obtain the embedding of a user $\hat{h}_u \in \mathbb{R}^d$ or an item $\hat{h}_i \in \mathbb{R}^d$ with a readout function on all the hidden representations from each layer:

$$\hat{h}_u^l = \sum_{j \in \mathcal{N}(u)} \hat{A}_{uj}\hat{h}_j^{l-1}, \quad \hat{h}_i^l = \sum_{j \in \mathcal{N}(i)} \hat{A}_{ij}\hat{h}_j^{l-1}$$
$$\hat{h}_u = \text{meanpooling}(\hat{h}_u^0, \hat{h}_u^1, ..., \hat{h}_u^{L_{ui}}), \quad \hat{h}_i = \text{meanpooling}(\hat{h}_i^0, \hat{h}_i^1, ..., \hat{h}_i^{L_{ui}}) \tag{6}$$

$\hat{h}_u^0 = \tilde{h}_u^0$ and $\hat{h}_i^0 = \tilde{h}_i^0$ denotes the ID embedding of user $u$ and item $i$, respectively.

Finally, we obtain the user representation $h_u$ through addition of $\tilde{h}_u$ and $\hat{h}_u$ and we can get the item representations $h_i$ in the same way:

$$h_u = \tilde{h}_u + \hat{h}_u, \quad h_i = \tilde{h}_i + \hat{h}_i \tag{7}$$

For Implicit Alignment, we adopt the Pairwise Bayesian Personalized Ranking Loss [12] to align the user representation with item representation, which can be defined as follows:

$$L_{IA} = \sum_{(u,i,j)\in D} -log(\sigma(h_u^T h_i - h_u^T h_j)) \tag{8}$$

where $\mathcal{D}$ is the set of training samples in which each triple $(u,i,j)$ satisfies $A_{ui} = 1$ and $A_{uj} = 0$, and $\sigma(\cdot)$ is the sigmoid function.

### 3.3 Explicit Alignment

**Multi-view Generator.** We use a simple Multi-View Generator to generate multi-views of representation. We feed the original embedding $h$ into a predictor of MLP to generate the online view. Meanwhile, we use the dropout technique to generate the masked view of the original embedding:

$$\overline{h} = hW_p + b_p, \quad \dot{h} = \mathbf{m} \odot h \tag{9}$$

where $W_p \in \mathbb{R}^{d\times d}$, $b_p \in \mathbb{R}^d$ denote the linear transformation matrix and bias. $\odot$ denotes the element-wise multiplication of two vectors. The mask vector $\mathbf{m} \in \{0,1\}^d$ is generated by sampling from a Bernoulli distribution. We place stop-gradient on the masked view $\dot{h}$.

**Multi-view Alignment.** For each uni-modal vector (text or vision), we use a linear transformation matrix to project the vector into a latent space.

$$h_i^m = x_i^m W_m + b_m, m \in \mathcal{M} \tag{10}$$

In this way, each uni-modal representation $h_i^m$ shares the same dimension with $h_i$. After we obtain the user representation $h_u$, item representation $h_i$, and uni-modal representations $h_i^m$, we can forward them into the Multi-View Generator for generating the masked view $(\dot{h}_u, \dot{h}_i, \dot{h}_i^v, \dot{h}_i^t)$ and the online view $(\overline{h}_u, \overline{h}_i, \overline{h}_i^v, \overline{h}_i^t)$.

In the explicit alignment process, we aim to achieve several objectives. Firstly, we further align user and item embeddings. We define a symmetric loss function using the negative cosine similarity between pairs $(\overline{h}_u, \dot{h}_i)$ and $(\overline{h}_i, \dot{h}_u)$:

$$L_{ui} = \sum_{(u,i)\in D} C(\overline{h}_u, \dot{h}_i) + C(\overline{h}_i, \dot{h}_u), \quad C(h_u, h_i) = -\frac{h_u^T h_i}{||h_u||_2 ||h_i||_2} \tag{11}$$

Specially, $\mathcal{D}$ is the set of positive training samples. We do not need negative samples when computing this loss. Secondly, to explicitly utilize the multimodal features, we further align the multimodal features of items with their target ID embeddings using negative cosine similarity loss. Additionally, for more robust learning of feature vectors, we add an intra-modality feature masked loss.

$$L_{im} = \sum_{m\in \mathcal{M}}\sum_{i\in \mathcal{I}} C(\overline{h}_i^m, \dot{h}_i), \quad L_{mm} = \sum_{m\in \mathcal{M}}\sum_{i\in \mathcal{I}} C(\overline{h}_i^m, \dot{h}_i^m) \tag{12}$$

All the above-mentioned losses are weighted and summed as follows:

$$L_{EA} = L_{ui} + \beta(L_{im} + L_{mm}) \tag{13}$$

### 3.4   Modality Feature Persistence

In prior works, multimodal features can be updated along with the optimization of the loss function. We experimentally find that the original information of the modalities would be lost during the training in the Sect. 4.6. Therefore, we use MSE Loss to constrain the update of multimodal features, which can ensure the persistence of multimodal information. That is, we freeze the original multimodal feature $(x_i^m)_F$ and learn an updatable copy $x_i^m$ during the training. The MFP loss is defined as:

$$L_{MFP} = \sum_{m \in \mathcal{M}} \sum_{i \in \mathcal{I}} ||x_i^m - (x_i^m)_F||_2 \tag{14}$$

Finally, our loss function can defined with hyperparameters $\lambda$, $\gamma$ as follows:

$$L = \lambda L_{IA} + (1 - \lambda)L_{EA} + \gamma L_{MFP} \tag{15}$$

## 4   Experiment

We perform comprehensive experiments to evaluate the effectiveness of DAMORE to answer the following questions.

- **RQ1**: How does DAMORE perform compared to the state-of-the-art multimodal recommendation models?
- **RQ2**: In the training paradigm with dual alignment, how does each alignment affect the results differently?
- **RQ3**: How does Modal Feature Persistence (MFP) maintain modal information and what impact does it have on model performance?

### 4.1   Experimental Datasets

Following previous studies [22–24], we choose three per-category datasets, i.e., Baby, Sports and Electronics for performance evaluation. The raw data of each dataset are pre-processed with a 5-core setting on both items and users, and the statistics of the 5-core filtered datasets are presented in Table 2.

**Table 2.** Statistics of the experimental datasets.

| Datasets | # Users | # Items | # Interactions | Sparsity |
|---|---|---|---|---|
| Baby | 19, 445 | 7, 050 | 160, 792 | 99.88% |
| Sports | 35, 598 | 18, 357 | 296, 337 | 99.95% |
| Electronics | 192, 403 | 63, 001 | 1, 689, 188 | 99.99% |

## 4.2 Baseline Methods

To demonstrate the effectiveness of DAMORE, we compare it with the following recommendation methods, including General models:BPR [12], LightGCN [4]; Explicit Utilization Model: VBPR [6], MMGCN [19], DualGNN [16], BM3 [24] and Implicit Utilization Model: GRCN [18], LATTICE [20], FREEDOM [23].

## 4.3 Setup and Evaluation Metrics

For a fair comparison, we follow the same evaluation setting of MMRec [22] with a random data splitting 8:1:1 on the interaction history of each user for training, validation, and testing. In the recommendation phase, all items that have not been interacted by the given user are regarded as candidate items. Moreover, we use Recall@K and NDCG@K to evaluate the top-K recommendation performance, where K at 10, 20 and we abbreviates the metrics of Recall@K and NDCG@K as R@K and N@K, respectively.

**Table 3.** Overall performance achieved by different recommendation methods in terms of Recall and NDCG. We mark the global best results on each dataset under each metric in boldface and the second best is underlined.

| Dataset | Metric | General Model | | Explicit Utilization Model | | | | Implicit Utilization Model | | | Ours |
| --- | --- | --- | --- | --- | --- | --- | --- | --- | --- | --- | --- |
| | | BPR | LightGCN | VBPR | MMGCN | DualGNN | BM3 | GRCN | LATTICE | FREEDOM | DAMORE |
| Baby | R@10 | 0.0357 | 0.0479 | 0.0423 | 0.0378 | 0.0448 | 0.0531 | 0.0539 | 0.0544 | 0.0626 | **0.0646** |
| | R@20 | 0.0575 | 0.0754 | 0.0663 | 0.0615 | 0.0716 | 0.0847 | 0.0833 | 0.0848 | 0.0992 | **0.0999** |
| | N@10 | 0.0192 | 0.0257 | 0.0223 | 0.0200 | 0.0240 | 0.0277 | 0.0288 | 0.0291 | 0.0330 | **0.0333** |
| | N@20 | 0.0249 | 0.0328 | 0.0284 | 0.0261 | 0.0309 | 0.0358 | 0.0363 | 0.0369 | 0.0420 | **0.0431** |
| Sports | R@10 | 0.0432 | 0.0569 | 0.0558 | 0.0370 | 0.0568 | 0.0635 | 0.0598 | 0.0618 | 0.0710 | **0.0725** |
| | R@20 | 0.0653 | 0.0864 | 0.0856 | 0.0605 | 0.0859 | 0.0962 | 0.0915 | 0.0947 | 0.1074 | **0.1097** |
| | N@10 | 0.0241 | 0.0311 | 0.0307 | 0.0193 | 0.0310 | 0.0341 | 0.0332 | 0.0337 | 0.0388 | **0.0394** |
| | N@20 | 0.0298 | 0.0387 | 0.0384 | 0.0254 | 0.0385 | 0.0425 | 0.0414 | 0.0422 | 0.0477 | **0.0490** |
| Elec | R@10 | 0.0235 | 0.0363 | 0.0293 | 0.0213 | 0.0363 | 0.0428 | 0.0389 | - | 0.0396 | **0.0463** |
| | R@20 | 0.0367 | 0.0540 | 0.0458 | 0.0343 | 0.0541 | 0.0639 | 0.0590 | - | 0.0601 | **0.0686** |
| | N@10 | 0.0127 | 0.0204 | 0.0159 | 0.0112 | 0.0202 | 0.0238 | 0.0216 | - | 0.0220 | **0.0260** |
| | N@20 | 0.0161 | 0.0250 | 0.0202 | 0.0146 | 0.0248 | 0.0292 | 0.0268 | - | 0.0273 | **0.0318** |

## 4.4 Performance Comparison

**Effectiveness (RQ1).** The performance from different methods on three datasets is summarized in Table 3. From the results, we have the following inclusion: In terms of all evaluation metrics including Recall and NDCG, DAMORE outperforms all baseline models on all datasets. In particular, the improvement achieved by DAMORE, based on the runner-up baseline, becomes more significant as data scale expands. This makes it more promising for application in industrial multimodal recommendation systems.

**Table 4.** Ablation study of DAMORE about dual alignment.

| Datasets | Variants | R@10 | R@20 | N@10 | N@20 |
|---|---|---|---|---|---|
| Baby | W/O Implicit | 0.0533 | 0.0826 | 0.0277 | 0.0352 |
|  | W/O Explicit | 0.0538 | 0.0852 | 0.0288 | 0.0369 |
|  | DAMORE | 0.0646 | 0.0999 | 0.0333 | 0.0431 |
| Sports | W/O Explicit | 0.0514 | 0.0823 | 0.0285 | 0.0364 |
|  | W/O Explicit | 0.0575 | 0.0894 | 0.0313 | 0.0395 |
|  | DAMORE | 0.0725 | 0.1097 | 0.0394 | 0.0490 |
| Elec | W/O Implicit | 0.0352 | 0.0545 | 0.0193 | 0.0242 |
|  | W/O Explicit | 0.0333 | 0.0511 | 0.0181 | 0.0227 |
|  | DAMORE | 0.0463 | 0.0686 | 0.0260 | 0.0318 |

### 4.5   Ablation Study

**The Effect of Dual Alignment (RQ2).** To investigate the impact of the dual alignment on model performance, we design the following variants of DAMORE: W/O Implicit and W/O Explicit. The W/O Implicit means $\lambda = 0$. In Table 4, we can observe that both alignments are crucial for the model to achieve good performance. When the implicit alignment is removed, the model's performance decreases significantly. We think that implicit alignment aims to preliminarily map user representations and item representations into the same semantic space. Based on that, explicit alignment uses multimodal features as additional supervision signals, further align item representation and multimodal features.

### 4.6   The Impact of Modal Feature Persistence(RQ3)

Since multimodal features are pre-trained on general data and may not fit in well with task-specific data naturally, previous works input them as learnable embeddings and update them in training. We argue that multimodal embeddings might be updated towards behavior signals and far from the original feature finally, causing the shift and loss of multimodal information. To confirm this, we conduct the experiment in Fig. 3(a) and track the change of the multimodal feature during the training process. According to the results, we find that the log of L2 distance between the updated and initial multimodal feature is increasing throughout the whole training process and has no sign of convergence. To tackle this problem, we propose MFP in Sect. 4.6. Results in Fig. 3(a) show that the L2 distance of multimodal features with MFP is growing but within a limited range and quickly converges, therefore keeping the original multimodal information reletively unlost.

Additionally, as shown in Fig. 3(b), we research model performance in different modal feature setting: W/O MFP(the modal feature can be updated), Freeze(the modal feature can't be updated), With MFP(the update of features is constrained by MFP). When we freeze multimodal features, the model performs poorly, which once again indicates a significant gap between original multimodal features and recommendation task. When allowing the multimodal features to be updated, the model's performance improves, but the corresponding multimodal features gradually lose the modality-specific information. After using MFP (Modal Feature Persistence), the model's performance further improves, and more importantly, multimodal features quickly converge to be nearly identical to the original modality features.

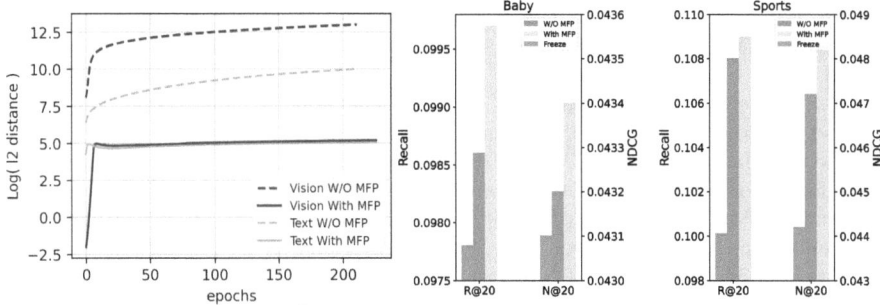

**Fig. 3.** (a) The performance drop of methods on Sports dataset when we randomly initialize multimodal features; (b) The model performance difference in different modal feature setting.

## 5  Conclusion

Explicitly and implicitly leveraging multimodal information is crucial for multimodal recommendation systems. Our work demonstrates the limitations of previous works in handling multimodal features of different scales. Furthermore, we successfully combine explicit and implicit utilization through a dual-aligned multimodal model DAMORE, which achieves state-of-the-art results on three public datasets. However, our work still presents several limitations: (i) In DAMORE, we have only explored one approach to combining explicit and implicit utilization. Further investigation is needed into deeper integration and the challenges within this combination. (ii) We employed a simple regularization technique to address the modal feature shift issue, and we intend to explore more effective strategies to resolve this problem in future research.

## References

1. Chen, X., He, K.: Exploring simple Siamese representation learning. In: Proceedings of the IEEE/CVF Conference on Computer Vision and Pattern Recognition, pp. 15750–15758 (2021)

2. Dosovitskiy, A., et al.: An image is worth 16x16 words: transformers for image recognition at scale. arXiv preprint arXiv:2010.11929 (2020)
3. Grill, J.B., et al.: Bootstrap your own latent-a new approach to self-supervised learning. In: Advances in Neural Information Processing Systems, vol. 33, pp. 21271–21284 (2020)
4. He, K., Fan, H., Wu, Y., Xie, S., Girshick, R.: Momentum contrast for unsupervised visual representation learning. In: Proceedings of the IEEE/CVF Conference on Computer Vision and Pattern Recognition, pp. 9729–9738 (2020)
5. He, K., Zhang, X., Ren, S., Sun, J.: Deep residual learning for image recognition. In: Proceedings of the IEEE Conference on Computer Vision and Pattern Recognition, pp. 770–778 (2016)
6. He, R., McAuley, J.: VBPR: visual Bayesian personalized ranking from implicit feedback. In: Proceedings of the AAAI Conference on Artificial Intelligence, vol. 30 (2016)
7. He, X., Deng, K., Wang, X., Li, Y., Zhang, Y., Wang, M.: LightGCN: simplifying and powering graph convolution network for recommendation. In: Proceedings of the 43rd International ACM SIGIR Conference on Research and Development in Information Retrieval, pp. 639–648 (2020)
8. Liu, Q., Wu, S., Wang, L.: DeepStyle: learning user preferences for visual recommendation. In: Proceedings of the 40th International ACM SIGIR Conference on Research and Development in Information Retrieval, pp. 841–844 (2017)
9. Liu, Y., et al.: AlignREC: aligning and training in multimodal recommendations (2024)
10. Radford, A., et al.: Learning transferable visual models from natural language supervision. In: International Conference on Machine Learning, pp. 8748–8763. PMLR (2021)
11. Radford, A., et al.: Language models are unsupervised multitask learners. OpenAI Blog $\mathbf{1}$(8), 9 (2019)
12. Rendle, S., Freudenthaler, C., Gantner, Z., Schmidt-Thieme, L.: BPR: Bayesian personalized ranking from implicit feedback. arXiv preprint arXiv:1205.2618 (2012)
13. Simonyan, K., Zisserman, A.: Very deep convolutional networks for large-scale image recognition. arXiv preprint arXiv:1409.1556 (2014)
14. Tao, Z., et al.: Self-supervised learning for multimedia recommendation. IEEE Trans. Multimedia (2022)
15. Vaswani, A., et al.: Attention is all you need. In: Advances in Neural Information Processing Systems, vol. 30 (2017)
16. Wang, Q., Wei, Y., Yin, J., Wu, J., Song, X., Nie, L.: DualGNN: dual graph neural network for multimedia recommendation. IEEE Trans. Multimedia (2021)
17. Wang, X., et al.: VIS+ AI: integrating visualization with artificial intelligence for efficient data analysis. Front. Comput. Sci. $\mathbf{17}$(6), 176709 (2023)
18. Wei, Y., Wang, X., Nie, L., He, X., Chua, T.S.: Graph-refined convolutional network for multimedia recommendation with implicit feedback. In: Proceedings of the 28th ACM MM, pp. 3541–3549 (2020)
19. Wei, Y., Wang, X., Nie, L., He, X., Hong, R., Chua, T.S.: MMGCN: multi-modal graph convolution network for personalized recommendation of micro-video. In: Proceedings of the 27th ACM MM, pp. 1437–1445 (2019)
20. Zhang, J., Zhu, Y., Liu, Q., Wu, S., Wang, S., Wang, L.: Mining latent structures for multimedia recommendation. In: Proceedings of the 29th ACM International Conference on Multimedia, pp. 3872–3880 (2021)
21. Zhang, K., et al.: Dream: a dual representation learning model for multimodal recommendation (2024)

22. Zhou, H., Zhou, X., Zeng, Z., Zhang, L., Shen, Z.: A comprehensive survey on multimodal recommender systems: taxonomy, evaluation, and future directions. arXiv preprint arXiv:2302.04473 (2023)
23. Zhou, X.: A tale of two graphs: freezing and denoising graph structures for multi-modal recommendation. arXiv preprint arXiv:2211.06924 (2022)
24. Zhou, X., et al.: Bootstrap latent representations for multi-modal recommendation. In: Proceedings of the ACM Web Conference 2023, pp. 845–854 (2023)

# CASINet: A Context-Aware Social Interaction Rumor Detection Network

Chang Yang[ID], Peng Zhang[(✉)], Hui Gao, and Jing Zhang

College of Intelligence and Computing, Tianjin University, Tianjin, China
{yangchang,pzhang,hui_gao,zhang_jing}@tju.edu.cn

**Abstract.** With the rapid expansion of social networks, the widespread propagation of rumors poses a significant threat to information security and societal stability. There are some challenges in current rumor detection work. Firstly, social media data's disorderly and chaotic nature presents substantial obstacles to extracting deep semantic features. Secondly, the diverse and complex user interaction relationships in social networks make it difficult to mine potential interaction features. To address these challenges, this paper introduces the Context-Aware Social Interaction Rumor Detection Network (CASINet), designed to explore contextual semantic features and social network user interaction characteristics, alongside the efficient fusion of heterogeneous information. The framework comprises three core components: The Contextual Semantic Interaction Module, which includes a context-aware semantic encoder and multi-level feature extractor for mining deep contextual semantic features; The Social Network User Interaction Module, which constructs a heterogeneous graph between users and tweets to capture the latent interaction relationships; and the Heterogeneous Feature Fusion Module, which enhances the model's generalization ability by automatically aligning and deeply integrating heterogeneous features. Experimental results validate the substantial enhancements achieved by our model in early rumor detection, providing a potent solution to the pervasive issue of rumor propagation in social networks.

**Keywords:** Rumor detection · Graph neural network · Feature fusion · Social networks · Natural language processing

## 1 Introduction

In the digital age, social media platforms greatly facilitate the rapid dissemination and sharing of information, while also providing fertile ground for rumors to spread rapidly. Rumors often appear under specific social psychological conditions, such as public fear of the unknown, curiosity about events, or anxiety in the face of uncertainty, prompting individuals to spread information without sufficient verification [14]. The decentralized and distributed nature of social networks further exacerbates the speed and scope of information dissemination, making it difficult for the public to distinguish truth from falsehood. During

X. He et al. (Eds.): CCIR 2024, LNCS 15418, pp. 28–40, 2025.
https://doi.org/10.1007/978-981-96-1710-4_3

the 2024 U.S. election, fraudulent robocalls impersonating presidential candidates misleadingly urged voters to abstain from primary voting under the false pretense of preserving their general election voting rights[1], underscoring the critical role of rumor detection in mitigating the spread of potentially impactful misinformation on public opinion. Due to the complex information dissemination environment, existing rumor detection models face some challenges. First, existing rumor detection methods are difficult to deal with the complexity and informality of discourse on social networks. For example, while a convolutional neural network (CNN) can effectively capture deep features, its max-pooling layer may result in the loss of detail and location information, thereby limiting the model's ability to comprehensively encode text sequences [15]. In addition, with the intricate user interactions on social networks, how to model this social interaction information and mine important information is a huge challenge to improve the accuracy of rumor detection.

To address these challenges, we developed the Context-Aware Social Interaction Rumor Detection Network (CASINet), which aims to improve the accuracy of rumor detection through deep semantic feature mining and effective modeling of user interactions. CASINet consists of three modules: the contextual semantic interaction module utilizes the contextual semantic encoder and multilevel feature extractor to comprehensively capture text information, improving the model's ability to handle the semantics and context of social media texts. Inspired by capsule networks [15], the dynamic routing mechanism replaces the traditional pooling operation to preserve more critical detailed features. In the social network user interaction module, a heterogeneous graph between users and tweets is constructed and a graph attention network is applied to effectively capture complex interactions within social networks. Finally, through the heterogeneous data fusion module, our model effectively integrates heterogeneous features, thereby improving accuracy and generalization. The contributions of this paper are as follows:

- We introduced the Context-Aware Social Interaction Rumor Detection Network (CASINet), which combines the Contextual Semantic Interaction Module and Social Network User Interaction Module. This innovation enhances the model's understanding of context in social media texts and highlights the key role of deep contextual semantic features and user interaction features in rumor detection.
- Based on the in-depth exploration of contextual semantic features and social network user interaction features, we also design an efficient heterogeneous information fusion strategy to automatically align and deeply fuse heterogeneous features, thereby enhancing the generalization ability of rumor detection.
- Experiments on two real-world datasets show that CASINet has superior performance, excels in early rumor detection tasks, and exhibits superior robustness.

---

[1] https://www.bbc.com/news/world-us-canada-68064247.

## 2   Related Work

In the nascent phase of rumor detection research, feature engineering emerged as a pivotal methodology. These approaches [1, 4, 8], leveraging multidimensional analyses of social media posts, user behavior patterns, and information dissemination paths, were instrumental in constructing effective supervised learning models aimed at accurately distinguishing between factual information and rumors. Castillo et al. [4] introduced a comprehensive framework assessing tweet credibility by integrating textual content, user characteristics, topic information, and propagation features. Lim et al. [8] developed an interactive framework that utilized social media users' opinions and feedback to validate the authenticity of information. Al-Zoubi et al. [1] employed a hybrid approach that combined Support Vector Machine (SVM) and K-Nearest Neighbors (K-NN) machine learning techniques, using tf-idf text features along with sentiment analysis to assess users' attitudes and stances toward tweets. While feature engineering methods have successfully captured the characteristics of social media information from various dimensions, the complexity, and diversity of social media data, compounded by the use of informal language, pose significant challenges to the complexity and generalizability of feature engineering.

Recently, the advancement of deep learning technologies has increasingly positioned automatic deep feature learning as a prevailing trend in rumor detection research. Initially, Ma et al. [10] pioneered the application of deep learning to rumor detection, employing Recurrent Neural Networks (RNNs) to capture the temporal evolution of context in related posts. Subsequently, RNNs and Convolutional Neural Networks (CNNs) [2, 9, 10, 21] were explored to autonomously extract latent features from text content and user behaviors. Additionally, models incorporating have been introduced to enhance the efficiency of learning temporal features of information and classification performance in rumor detection [9]. More recently, attention mechanisms [5] and pre-trained models [18, 19] have been applied in this domain.

With the evolution of Graph Neural Networks (GNNs), the ability to capture contextual relationships, syntactic structures, and semantic dependencies within text has led to their application in the Natural Language Processing (NLP) domain. The advent of GNNs introduced a new perspective to rumor detection, especially in analyzing information propagation structures on social media. By modeling them as propagation trees or graphs, GNNs excel at capturing the complex relationships and dynamic changes occurring during information dissemination. Ma et al. [12] proposed two recursive neural models leveraging both bottom-up and top-down tree-structured neural networks to capture the hidden representations of all recursive content. Yuan et al. [22] developed the Global-Local Attention Network (GLAN) to describe the relationships between tweets, retweets, and users as heterogeneous graphs, thus capturing local semantic information and global structural information relevant to rumors. Dou et al. [6] introduced a graph-structured representation framework for modeling participants and their interactions within social networks, capturing patterns of social structure involvement. Yan et al. [17] proposed a heterogeneous graph attention network with a bidirectional information propagation mechanism for rumor detection, effectively modeling users, tweets, and their interactions to simulate the dynamics of social network participants and interactions.

# 3 Problem Statement

In this paper, we formulate the rumor detection problem as a classification task. Specifically, we aim to devise a function $p(Y|V, G; \theta)$ that predicts the probability of an event belonging to one of four possible categories, with all learnable model parameters denoted by $\theta$. These categories are true rumors (T), false rumors (F), non-rumors (N), and unverified rumors (U).

For semantic content features $V$, our focus is on the collection of textual information encompassing the original tweet and its associated comments. For each specific event $X = \{x, c_1, c_2, ..., c_p\}$, $x$ represents the source tweet, while $c_i$ denotes the $i^{th}$ comment associated with the source tweet, and $p$ signifies the maximum number of comments. Furthermore, to gain a deeper understanding of the patterns of user interactions, we construct a heterogeneous graph of user-tweet interactions $G_i = (V_i, E_i)$ for each event, representing the interaction structure features in a graphical form. $V_i$ and $E_i$ represent the nodes and edges of the graph, respectively. The nodes consist of the source tweet $x$ and the set of users $U_i = \{u_1, u_2, ..., u_n\}$ who have directly interacted with the tweet. The edges are weighted inversely to the time of user retweets or replies to the original tweet, reflecting the interaction relationship between users and the source tweet.

# 4 Methodology

The model we proposed, Context-Aware Social Interaction Rumor Detection Network(CASINet), is illustrated in Fig. 1 and mainly contains three modules, namely Contextual Semantic Interaction Module, Social Network User Interaction Module, and Heterogeneous Feature Fusion Module. The subsequent sections provide a detailed explanation of the principles and specifics of each module.

## 4.1 Contextual Semantic Interaction Module

To address the challenge of effectively extracting and understanding deep semantic features from widely dispersed and complex content on social media for rumor detection tasks, we first integrate BERTweet [13], pre-trained on massive social media text data, as a contextual semantic encoder. Furthermore, we combine the Bidirectional Long Short-Term Memory Network (BiLSTM) and Capsule Network to perform multi-level semantic extraction, aiming to more comprehensively capture the basic information of the text. It is worth noting that the introduction of the dynamic routing mechanism replaces the traditional pooling operation, preventing the loss of details and location information, and enhancing the model's ability to capture and retain key information in the text.

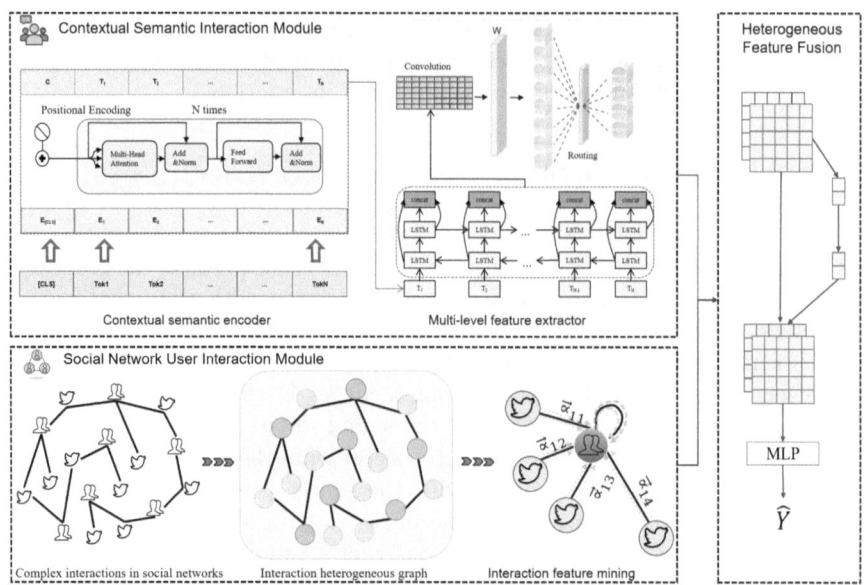

**Fig. 1.** Context-Aware Social Interaction Rumor Detection Network

Specifically, we consider an event consisting of a source tweet $x_i$ and a series of related comments $C_i = \{c_1, c_2, \ldots, c_m\}$ where $m$ represents the maximum number of comments related to the source tweet. These textual sequences are fed into BERTweet $H_i^B = \text{BERTweet}(x_i \cup C_i)$.

To further delve into sequential and hierarchical feature learning of social media texts, we employ the BiLSTM as the sequential feature learning layer. The BiLSTM efficiently captures long-distance dependencies within text data, with its bidirectional architecture considering both forward and backward contextual information for a comprehensive semantic representation $H_i^{\text{BiLSTM}}$. At the hierarchical feature learning stage, a 1D convolutional layer first processes the output of the BiLSTM, to extract n-gram features from various positions.

$$H_i^{\text{BiLSTM}} = [\overrightarrow{LSTM}(H_i^B), \overleftarrow{LSTM}(H_i^B)] \in \mathbb{R}^{s \times 2h} \tag{1}$$

$$u_i = Conv(H_i^{BiLSTM}) \tag{2}$$

where $u_i \in \mathbb{R}^{(s-k+1) \times k\_num}$ represents the processed capsule representation post-convolution, with $k$ and $k\_num$ denoting the size and number of convolutional kernels, respectively. The "Squash" non-linear activation function then normalizes $u_i$, maintaining the proportionality between the output vector length of each lower-level capsule and its significance in the higher-level capsules.

$$u_i = Squash(u_i) = \frac{||u_i||}{1 + ||u_i||^2} \frac{u_i}{||u_i||} \tag{3}$$

Subsequently, the prediction vectors for the lower-level capsules are obtained through projection, with the high-level capsules' prediction vectors being the weighted sum of these prediction vectors and their coupling coefficients.

$$v_k = \sum_i c_{k|i} \cdot \hat{u}_{k|i}, \tag{4}$$

$$c_{k|i} = \text{softmax}(b_{k|i}), \tag{5}$$

$$b_{k|i} = b_{k|i} + \hat{u}_{k|i} \cdot v_k \tag{6}$$

where $b_{k|i}$ is initially set to zero and updated through iterations, representing the coupling coefficients of lower-level capsules, which determines the contribution of the low-level vector to the high-level vector. Through iterative dynamic routing, we derive the final output of the capsule convolutional layer, $V_i = [v_0, v_1, ..., v_p]$, with each $v_k \in \mathbb{R}^{c_{in} \times c_{\text{dim}}}$ representing the semantic content information of the event, encapsulating both depth and breadth of the contextual and interactive features inherent in social media texts.

### 4.2 Social Network User Interaction Module

In social media platforms, the interactions among users and between users and information often harbor crucial clues about the veracity of information. To effectively capture these latent relationships, we model user interactions on social networks by constructing a user-tweet heterogeneous graph and employ a Graph Attention Network (GAT), dynamically learning the influence among nodes. As iterations progress, nodes increasingly gather information from their multi-hop neighbors, culminating in a comprehensive global representation of each node. Recognizing the varying significance of different nodes within social networks, GAT employs concatenation and normalization operations to yield asymmetric attention scores. This asymmetric allocation of attention is instrumental in precisely identifying and leveraging key nodes-such as influential users or critical information dissemination paths-thus enhancing the accuracy and efficiency of rumor detection. Specifically, the interaction heterogeneous graph $G_i^{interact}$ consists of node $s_i$ and a set of neighbor nodes $U_i = \{u_1, u_2, ..., u_n\}$. GAT computes attention coefficients $\alpha$ for the neighbors of each $s_i$, combining these neighbors and their respective attention coefficients to obtain the final output representation:

$$\alpha_{ij} = \frac{\exp\left(\text{LeakyReLU}\left(a^T[W_s s_i \| W_u u_j]\right)\right)}{\sum_{k \in N(s_i)} \exp\left(\text{LeakyReLU}\left(a^T[W_s s_i \| W_u u_k]\right)\right)} \tag{7}$$

$$I_i = \sigma\left(\sum_{j \in N(s_i)} \alpha_{ij} W u_j\right) \tag{8}$$

where $\sigma$ is a non-linear function, $a^T$ represents the weights of the neural network, and $N(s_i)$ denotes the neighbors of node $s_i$, including itself. Through this mechanism, GAT filters out noisy neighbors, enhancing model performance and facilitating specific interpretations of the results.

### 4.3   Heterogeneous Feature Fusion Module

Finally, this module is tasked with integrating semantic content features $V$, and interaction structure features $I$, overcoming the limitations of simplistic concatenation strategies that disrupt the intrinsic linkage between semantic content and user interaction information. Given the distinct learning methodologies for semantic and interaction features, merging them directly within a single vector space is impractical. Drawing inspiration from channel attention mechanisms in computer vision, we propose a sophisticated method for feature fusion.

For each event, we transform both semantic and interaction features into a common space while preserving their original spatial relationships. Transformed semantic content features $V' = W_v V$ and transformed interaction structure features $I' = W_i I$ serve as input, where $W_v$ and $W_i$ are transformation functions that map the original features into a unified space while preserving their spatial relationships. This process is akin to treating features as multiple channels in a color image. The fusion representation is derived through a channel attention framework, beginning with feature compression via averaging operations to initialize attention scalars for each feature channel.

$$z_i = F_{sq}(I) = \text{Avg}(I) \tag{9}$$
$$z_v = F_{sq}(V) = \text{Avg}(V) \tag{10}$$

Feature expansion follows, executed through dual fully connected layers that capture complex inter-channel relationships

$$s_i = F_{ex}(z_i, W) = \sigma(W_2\delta(W_1 z_i)) \tag{11}$$
$$s_v = F_{ex}(z_v, W) = \sigma(W_2\delta(W_1 z_v)) \tag{12}$$

The first layer, $W_1$, reduces dimensionality to elucidate channel dependencies, while the second layer, $W_2$, reinstates the feature dimensions to preserve the richness of information. $\sigma$ and $\delta$ represent the *Sigmoid* and *Relu* activation functions respectively. Ultimately, the fusion representation, $E$, is formed by combining the weighted feature maps with the original ones through

$$E = F_{scale}(I, s_i) + F_{scale}(V, s_v) = s_i I + s_v V \tag{13}$$

This method not only contemplates the intricate interactions between features but also dynamically adjusts the contribution of each feature, thereby enhancing the model's capacity for rumor detection.

## 5   Experiments

### 5.1   Dataset and Experimental Setup

To evaluate our proposed model for rumor detection, we selected two widely recognized public datasets in the field, Twitter15 and Twitter16 [11]. These datasets encompass four categories of rumors: true rumors (TR, false rumors (F), unverified rumors (U), and non-rumors (N).

For the experimental setup, our model was implemented using the PyTorch framework and computations were performed on an NVIDIA Tesla V100 GPU. We divided the dataset into training, validation, and testing sets with proportions of 70%, 20%, and 10%, respectively, to ensure extensive utilization of data in model evaluation. During training, we configured the batch size to 64 and set the number of epochs to 30, with an initial learning rate of 1e-5. Additionally, L2 regularization with a weight of 0.001 was employed to manage model complexity. The training phase utilized the cross-entropy loss function to quantify the discrepancy between predicted outcomes and actual labels. Moreover, an early stopping mechanism [20] was applied to further prevent overfitting.

## 5.2 Baselines

In experiments, we select a series of baseline models that have proven superior performance in natural language processing (NLP) and are widely used in text classification and rumor detection. 1) BERT: A state-of-the-art pre-trained model using Transformer encoders to effectively comprehend complex language contexts. 2) FastText: An efficient text classification tool that utilizes word and n-gram features for rapid and effective training. 3) TextCNN: Employs convolutional neural networks to extract deep features for text classification tasks. 4) TextRNN: Utilizes recurrent neural networks to process sequential data, capturing text's temporal dynamics. 5) dEFEND [16]: Uses attention mechanisms to identify semantic correlations between tweets and comments, thereby improving rumor detection. 6) RvNN [12]: Employs Recurrent Neural Networks for analyzing propagation trees, aimed at effective rumor identification. 7) BiGCN [3]: Leverages graph convolutional networks on bidirectional propagation graphs for enhanced rumor detection. 8) STS-NN [7]: A spatio-temporal structured neural network that combines spatial analysis of tree structures, temporal understanding through GRUs, and feature integration via self-attention mechanisms for advanced rumor detection.

## 5.3 Experimental Results

In this section, we compare the performance of our proposed CASINet with a series of baseline models, as shown in Table 1, which include models designed for the text classification task (BERT, FastText, TextCNN, and TextRNN) and customized for the rumor detection task models (dEFEND, RvNN, BiGCN, and STS-NN).

In the comparison of text classification models, first, the contextual semantic encoder provides more powerful semantic encoding capabilities. Compared with TextCNN and TextRNN, the capsule network effectively solves the problem of loss of details and location information encountered in text data processing; and BERTweet digs deeper into the complex semantic relationships inherent in social network texts than BERT and FastText. Moreover, the implementation of

**Table 1.** Performance comparison on Twitter15 and Twitter16 datasets

| Model | Twitter15 | | | | | Twitter16 | | | | |
|---|---|---|---|---|---|---|---|---|---|---|
| | Acc | N | F | T | U | Acc | N | F | T | U |
| BERT | 0.761 | 0.731 | 0.746 | 0.817 | 0.768 | 0.742 | 0.649 | 0.703 | 0.811 | 0.737 |
| FastText | 0.722 | 0.630 | 0.749 | 0.724 | 0.729 | 0.730 | 0.684 | 0.709 | 0.765 | 0.674 |
| TextCNN | 0.728 | 0.716 | 0.658 | 0.743 | 0.776 | 0.703 | 0.692 | 0.688 | 0.786 | 0.614 |
| TextRNN | 0.703 | 0.697 | 0.711 | 0.656 | 0.703 | 0.712 | 0.709 | 0.714 | 0.773 | 0.647 |
| DEFEND | 0.731 | 0.631 | 0.646 | 0.617 | 0.668 | 0.721 | 0.649 | 0.603 | 0.611 | 0.637 |
| RvNN | 0.713 | 0.702 | 0.691 | 0.746 | 0.654 | 0.737 | 0.662 | 0.743 | 0.801 | 0.768 |
| BiGCN | 0.798 | 0.716 | 0.758 | 0.843 | 0.876 | 0.803 | 0.792 | 0.788 | 0.796 | 0.814 |
| STS-NN | 0.809 | 0.797 | 0.811 | 0.856 | 0.773 | 0.821 | 0.739 | 0.814 | 0.883 | 0.847 |
| **Ours** | **0.831** | **0.808** | **0.792** | **0.851** | **0.849** | **0.824** | **0.768** | **0.823** | **0.794** | **0.895** |

multi-level feature extractors allows our model to meticulously distill essential information from texts. This approach not only bolsters the flexibility and efficacy of the feature extraction process but also surmounts the constraints inherent in single-scale feature extraction methodologies exemplified by FastText.

In the comparative analysis with the rumor detection model, first, the model dEFEND, which mainly focuses on text content, has shortcomings in extracting deep semantic features. Conversely, CASINet demonstrates superior efficacy in leveraging structural feature-based models. Notably, BiGCN elucidates critical structural features of information dissemination using graph convolutional networks (GCN). CASINet's Social Network User Interaction Module establishes a heterogeneous graph between users and tweets, employing the Graph Attention Network (GAT) to foster a deeper comprehension of user interactions. Furthermore, STS-NN facilitates feature fusion via the self-attention mechanism. The heterogeneous feature fusion module in CASINet employs an effective fusion strategy that ensures the alignment and integration of disparate features, culminating in significant enhancements in the precision and efficiency of rumor detection.

### 5.4   Ablation Experiments

We conducted a series of ablation studies to assess the contribution of each component within the CASINet model. 1) W/o BERTweet: Replace the BERTweet component in the contextual semantic interaction module with standard BERT. 2) W/o Capsule Network: Remove the capsule network component in the contextual semantic interaction module. 3) W/o BiLSTM: Remove the BiLSTM component in the contextual semantic interaction module. 4) W/o SNUIM: Remove the Social Network User Interaction Module. 5) W/o HFFM: replace the Heterogeneous Feature Fusion Module with a simple feature connection.

**Table 2.** Accuracy of different models on Twitter15 and Twitter16 datasets.

| Model | Twitter15 Acc | Twitter16 Acc |
|---|---|---|
| Full model | 0.831 | 0.824 |
| W/o BERTweet | 0.796 | 0.784 |
| W/o Capsule Network | 0.813 | 0.802 |
| W/o BiLSTM | 0.817 | 0.811 |
| W/o SNUIM | 0.804 | 0.797 |
| W/o HFFM | 0.819 | 0.816 |

The results of the ablation experiments in Table 2 reveal the significant impact of each module on model performance. Specifically, first, replacing BERTweet with standard BERT resulted in the most significant performance degradation, reaching 3.5% and 4% respectively, emphasizing the superiority of BERTweet in capturing the semantic information of social media texts and processing text contextual relevance. Secondly, the removal of the capsule network and BiLSTM components also leads to performance degradation. These components play an important role in the model's understanding of complex text structure and semantics by aggregating input information from different dimensions and extracting rich feature representations and high-level semantic information. At the same time, our experiments also found that removing the social network user interaction module will lead to performance degradation. This module effectively enhances the cohesion between similar feature nodes by simulating the interaction between users and the relationship between users and content, thereby improving the performance of the model. Finally, the decrease in accuracy caused by replacing our heterogeneous feature fusion module with a simple splicing strategy further confirms the effectiveness of our fusion strategy in integrating multi-type features and improving model generalization ability and accuracy.

### 5.5   Early Rumor Detection

In the field of rumor detection, being able to promptly identify and contain rumors at an early stage is a key goal. Our study evaluates the performance of the model to delve into early rumor detection by controlling the gradual increase of comments.

As shown in Fig. 2, the experimental results demonstrate the superiority of our model over other baseline models. Especially in the early stages when reviews were limited, CASINet also showed good performance. BiGCN and RvNN often show instability in this case. Because these models rely on abundant data to discern complex relationships between nodes and edge networks, the lack of data hinders their ability to learn these relationships effectively. The results highlight our model's effectiveness in early rumor detection and the importance

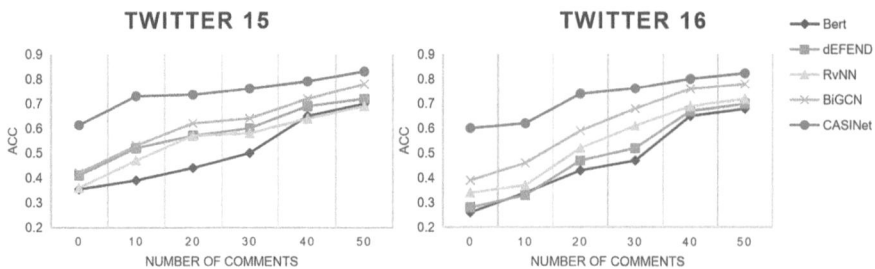

**Fig. 2.** Early rumor detection experimental results

of high-level semantic understanding and meticulous user interaction modeling as features of effective early rumor detection strategies.

# 6   Conclusion

This paper solves the key challenge of rumor detection on social media platforms by introducing an innovative deep learning architecture Context-Aware Social Interactive Rumor Detection Network (CASINet). This model aims to effectively identify and mitigate the spread of rumors by complexly analyzing rumor propagation mechanisms and user interaction patterns on social media. The novelty of CASINet lies in its ability to effectively align and fuse heterogeneous features, deep semantic features of text context, and structural features of social interactions between users, thereby enhancing rumor detection generalization capabilities. Experimental evaluation demonstrates the superiority of CASINet in mining complex semantic relationships, capturing multi-faceted features, modeling complex interactions between users and text, and providing superior text representation. Even with limited training data, its performance outperforms multiple existing baseline models. These findings verify the effectiveness of our model design, while also emphasizing the key role of effective fusion of heterogeneous features in rumor detection, providing new directions for future research to explore.

**Acknowledgement.** This work is supported in part by the Natural Science Foundation of China (grant No. 62276188). The authors have no competing interests to declare that are relevant to the content of this article.

**Disclosure of Interests.** The authors have no competing interests to declare that are relevant to the content of this article.

# References

1. Al-Ghadir, A.I., Azmi, A.M., Hussain, A.: A novel approach to stance detection in social media tweets by fusing ranked lists and sentiments. Inf. Fusion **67**, 29–40 (2021)
2. Asghar, M.Z., Habib, A., Habib, A., Khan, A., Ali, R., Khattak, A.: Exploringdeep neural networks for rumor detection. J. Ambient. Intell. Humaniz. Comput. **12**, 4315–4333 (2021)
3. Bian, T., et al.: Rumordetection on social media with bi-directional graph convolutional networks. In: Proceedings of the AAAI Conference on Artificial Intelligence, vol. 34, pp. 549–556 (2020)
4. Castillo, C., Mendoza, M., Poblete, B.: Information credibility on Twitter. In: Proceedings of the 20th International Conference on World Wide Web, pp. 675–684 (2011)
5. Chen, T., Li, X., Yin, H., Zhang, J.: Call attention to rumors: deep attention based recurrent neural networks for early rumor detection. In: Ganji, M., Rashidi, L., Fung, B.C.M., Wang, C. (eds.) PAKDD 2018. LNCS (LNAI), vol. 11154, pp. 40–52. Springer, Cham (2018). https://doi.org/10.1007/978-3-030-04503-6_4
6. Dou, Y., Shu, K., Xia, C., Yu, P.S., Sun, L.: User preference-aware fake news detection. In: Proceedings of the 44th International ACM SIGIR Conference on Research and Development in Information Retrieval, pp. 2051–2055 (2021)
7. Huang, Q., Zhou, C., Wu, J., Liu, L., Wang, B.: Deep spatial-temporal structure learning for rumor detection on Twitter. Neural Comput. Appl. **35**(18), 12995–13005 (2023)
8. Lim, W.Y., Lee, M.L., Hsu, W.: iFact: an interactive framework to assess claims from tweets. In: Proceedings of the 2017 ACM on Conference on Information and Knowledge Management, pp. 787–796 (2017)
9. Liu, Y., Wu, Y.F.: Early detection of fake news on social media through propagation path classification with recurrent and convolutional networks. In: Proceedings of the AAAI Conference on Artificial Intelligence, vol. 32 (2018)
10. Ma, J., et al.: Detecting rumors from microblogs with recurrent neural networks (2016)
11. Ma, J., Gao, W., Wong, K.F.: Detect rumors in microblog posts using propagation structure via kernel learning. Association for Computational Linguistics (2017)
12. Ma, J., Gao, W., Wong, K.F.: Rumor detection on twitter with tree-structured recursive neural networks. Association for Computational Linguistics (2018)
13. Nguyen, D.Q., Vu, T., Nguyen, A.T.: Bertweet: a pre-trained language model for English tweets. In: Proceedings of the 2020 Conference on Empirical Methods in Natural Language Processing: System Demonstrations, pp. 9–14 (2020)
14. Pröllochs, N., Feuerriegel, S.: Mechanisms of true and false rumor sharing in social media: collective intelligence or herd behavior? Proc. ACM Hum. Comput. Interact. **7**(CSCW2), 1–38 (2023)
15. Sabour, S., Frosst, N., Hinton, G.E.: Dynamic routing between capsules. In: Advances in Neural Information Processing Systems, vol. 30 (2017)
16. Shu, K., Cui, L., Wang, S., Lee, D., Liu, H.: Defend: explainable fake news detection. In: Proceedings of the 25th ACM SIGKDD International Conference on Knowledge Discovery and Data Mining, pp. 395–405 (2019)
17. Yan, M., Yang, W., Sun, B., Zhu, Y.: Heterogeneous graph attention networks with bi-directional information propagation for rumor detection. In: 2022 7th International Conference on Big Data Analytics (ICBDA), pp. 236–242. IEEE (2022)

18. Yang, C., Yu, X., Wu, J., Zhang, B., Yang, H.: Graph-aware multi-feature interacting network for explainable rumor detection on social network. Expert Syst. Appl. **249**, 123687 (2024)

19. Yang, C., Zhang, P., Qiao, W., Gao, H., Zhao, J.: Rumor detection on social media with crowd intelligence and ChatGPT-assisted networks. In: Proceedings of the 2023 Conference on Empirical Methods in Natural Language Processing, pp. 5705–5717 (2023)

20. Yao, Y., Rosasco, L., Caponnetto, A.: On early stopping in gradient descent learning. Constr. Approx. **26**, 289–315 (2007)

21. Yu, F., Liu, Q., Wu, S., Wang, L., Tan, T., et al.: A convolutional approach form is information identification. In: IJCAI, pp. 3901–3907 (2017)

22. Yuan, C., Ma, Q., Zhou, W., Han, J., Hu, S.: Jointly embedding the local and global relations of heterogeneous graph for rumor detection. In: 2019 IEEE International Conference on Data Mining (ICDM), pp. 796–805. IEEE (2019)

# A Claim Decomposition Benchmark for Long-Form Answer Verification

Zhihao Zhang, Yixing Fan$^{(\boxtimes)}$, Ruqing Zhang, and Jiafeng Guo

Institute of Computing Technology, Chinese Academy of Sciences, Beijing, China
{zhangzhihao22s,fanyixing,zhangruqing,guojiafeng}@ict.ac.cn

**Abstract.** The advancement of large language models (LLMs) has significantly boosted the performance of complex long-form question answering tasks. However, one prominent issue of LLMs is the generated "hallucination" responses that are not factual. Consequently, attribution for each claim in responses becomes a common solution to improve the factuality and verifiability. Existing researches mainly focus on how to provide accurate citations for the response, which largely overlook the importance of identifying the claims or statements for each response. To bridge this gap, we introduce a new claim decomposition benchmark, which requires building system that can identify atomic and checkworthy claims for LLM responses. Specifically, we present the Chinese Atomic Claim Decomposition Dataset (CACDD), which builds on the WebCPM dataset with additional expert annotations to ensure high data quality. The CACDD encompasses a collection of 500 human-annotated question-answer pairs, including a total of 4956 atomic claims. We further propose a new pipeline for human annotation and describe the challenges of this task. In addition, we provide experiment results on zero-shot, few-shot and fine-tuned LLMs as baselines. The results show that the claim decomposition is highly challenging and requires further explorations. All code and data are publicly available (https://github.com/FBzzh/CACDD).

**Keywords:** Claim Decomposition · Chinese Datasets · Large Language Model

## 1 Introduction

In recent years, LLMs have demonstrated excellent performance in various domains of Natural Language Processing (NLP) due to their robust natural language capabilities. Through training on extensive data with a large number of parameters, these models have shown significant advancements in their ability to understand and process natural language. One of the most notable improvements in LLMs is their ability for complex reasoning and the generation of lengthy, coherent responses, which has greatly enhanced their capability to answer complex, long-form questions. However, the powerful generation capabilities of these models have also led to the emergence of the so-called "hallucination" problem, where the models occasionally generate content that does not

© The Author(s), under exclusive license to Springer Nature Singapore Pte Ltd. 2025
X. He et al. (Eds.): CCIR 2024, LNCS 15418, pp. 41–53, 2025.
https://doi.org/10.1007/978-981-96-1710-4_4

align with reality. This issue not only affects the reliability of the models but also significantly limits their use in real-world applications, especially in high-stakes, risk-sensitive tasks where the accuracy of factual information is crucial.

In response to the issue of hallucination generation in large language models, researchers have proposed a variety of solutions to verify the credibility of the generated response, including multi-path result cross-validation, context consistency verification, and external knowledge verification. Initially, multi-path result cross-validation identifies and reduces potential hallucinations by comparing the output of different models or different parameter settings of the same model. For instance, Wang et al. [17] proposed the Self-Consistency method, which selects the most consistent answer by sampling different reasoning paths. Subsequently, context consistency verification focuses on ensuring that the text generated by the model is logically and semantically consistent with its input context, thereby reducing the generation of hallucinations. For example, Shi et al. [16] introduced Context-Aware Decoding (CAD), which amplifies the probability difference of the model's output with and without context, making the model more context-compliant. Lastly, external knowledge verification enhances the factual accuracy of the generated responses by comparing them with external knowledge bases or retrieved evidence documents. For instance, Gao et al. [5] proposed the Retrofit Attribution using Research and Revision (RARR), which reduces hallucinations in large model responses by detecting and modifying content inconsistent with retrieved relevant evidence.

Despite the fact that current research has proposed a variety of answer verification approaches from different perspectives, these approaches mainly focus on verifying the consistency between the response and the facts, often neglecting the importance of identifying and decomposing the claims that are worthy to be verified in the results. For instance, Yue et al. [21] regards the entire response as a single claim and compares it with the reference. While Gao et al. [6] assume that each natural sentence within the response constitutes an independent claim, and use the NLTK tool to identify sentences, subsequently comparing each sentence to the reference crafted by human experts. These approaches have not delved into the intrinsic nature of the claims, thereby constraining the accuracy and reliability of the answer verification.

In reality, the factual verification of long response is an extremely challenging task. This is primarily due to the fact that a lengthy response typically encompasses one or more claims, with complex co-reference relationships existing among these claims. As shown in Fig. 1, the answer to the question "Why is Stephen Hawking so important?" includes three claims, and there are referential relationships between claims 2 and 3 and claim 1. Additionally, the sentence "He is the greatest scientist in my mind" in the answer does not contain valuable information and does not require verification. It is evident that the verification of responses from LLMs usually requires: 1) identifying claims; 2) atomic decomposition of claims; 3) factual verification of each claim. Consequently, accurately identifying the claims in the response and decomposing each claim into atomic claims is an important component in the effective answer verification.

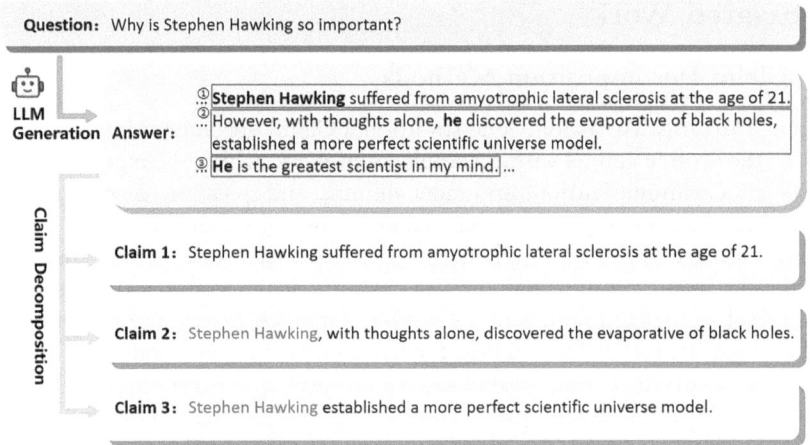

**Fig. 1.** A example of the claim decomposition.

To address the tasks mentioned, this paper introduces the first benchmark for Chinese claim decomposition in answer verification, aiming to enhance the reliability of LLMs in answering complex questions. Initially, we define the concept of atomic claims, inspired by Russell's philosophy of logical atomism [15], which encompasses four fundamental principles: Indivisibility, Semantic Integrity, Verifiability, and Context Independence. Subsequently, we design an annotation pipeline for atomic claim decomposition. Based on this pipeline, we have created the CACDD dataset through manual annotation, which is built on the WebCPM dataset. Each claim in this dataset has been resolved for co-references to satisfy the principle of context independence. In total, the CACDD includes 500 questions and 4956 claims, with all the annotated data and the code publicly accessible on https://github.com/FBzzh/CACDD.

In order to better understand the claim decomposition task, we have conducted experimental analyses to assess the performance of LLMs on this benchmark. Specifically, we have conducted zero-shot and few-shot experiments on open-source LLMs, as well as GPT-3.5. The results indicate that the open-source LLMs do not perform satisfactorily on this task, and even the outstanding GPT-3.5 still has a significant gap compared to humans.

The main contributions of this study include:

1. We have constructed the first Chinese Atomic Claim Decomposition Dataset (CACDD) for claim decomposition task, providing a valuable resource for future research;
2. We have tested the performance of the claim decomposition task on several advanced LLMs, providing a foundation for further optimization;
3. We have provided a clear definition of atomic claims, offering a standard that can be followed for the claim decomposition task.

## 2    Related Work

### 2.1    Claim Decomposition Methods

**Lexical Parsing Methods.** The traditional Claim decomposition task aims to identify the atomic claims within complex sentences, which is commonly used for tasks such as summarization, argument mining, and question answering. Before the advent of large language models, most related research was based on lexical parsing methods. The DCP [7] method extracts the verb phrase components and clauses from complex sentences as criteria for summarization tasks. The DisSim [13] framework transforms complex sentences into a more regular intermediate representation by dividing them into a two-layered semantic hierarchy consisting of core facts and accompanying context. Moreover, the PredPatt [20] framework decomposes complex sentences by extracting predicate argument structures from syntactic dependency parsing based on the Universal Dependencies [4] Project. However, these methods based on lexical parsing typically only generate a structured representation of claims, rather than complete and coherent sub-claims, thereby constraining their application in downstream NLP tasks such as fact-checking and answer verification.

**LLM Prompting Approaches.** Upon their emergence, LLMs have garnered widespread attention across various domains of natural language processing due to their outstanding natural language comprehension and generation capacities. Recent research pertaining to claim decomposition tasks has also capitalized on the language ability of LLMs. Most of these LLM-based approaches are implemented through prompts [3,9,10,12,18]. By providing examples that are carefully crafted by humans, they instruct LLMs to complete this task through in-context learning. In contrast to lexical parsing methods, these LLM prompting approaches are more flexible and unstructured, enabling the generation of coherent and fluent claims. Nonetheless, the LLM prompting approaches are also affected by the hallucinations inherent in LLMs, which can result in the generation of sub-claims that deviate from the original text. Furthermore, to ensure the quality of claim decomposition, researchers are required to provide enough in-context examples, leading to the prompts often being excessively lengthy. The excessive length of these prompts can lead to a reduction in inference efficiency and an increase in token costs.

### 2.2    Answer Verification

The verification of the extent to which LLM-generated responses are supported by the context or retrieved evidence is highly dependent on the claim decomposition task. FactScore [12] evaluates the factual precision of the LLM-generated responses through decompose them into sub-claims and verify each claim using a knowledge source. Factcheck-GPT [18] introduces an end-to-end annotation framework that assesses the factual accuracy of LLM-generated responses through a combined approach that involves both LLM and human evaluation. Chen [2] employs problem decomposition to decompose LLM-generated responses, verifying complex claims from various aspects, both explicit and implicit, by generating a series of yes or no sub-questions. However, the datasets for these works are primarily derived from fact-intensive domains such as biographies [12] and political claims [1,2], exhibiting a deficiency in data and pertinent research from long-form question-answering scenarios.

# 3   The Atomic Claim Decomposition Task

In this section, we first introduce the definition of atomic claim along with its respective characteristics, followed by an exposition of the definition and challenges associated with the claim decomposition task.

## 3.1   Atomic Claim Definition

The definition of atomic of Claims has long been a contentious issue. Wang [18] believes that it is challenging to define atomic and then determine the granularity of decomposition. Unfortunately, their discourse on atomic did not arrival at a definitive conclusion. Wanner [19] traces this issue back to the philosophical level. They employ Bertrand Russell's philosophy of logical atomism [15] and the neo-Davidsonian analysis [11] as the theoretical guidance, and manually decomposed 21 examples as the guidance for LLMs to perform the claim decomposition task.

We also fellow the guidance of Bertrand Russell's philosophy of logical atomism and provide the following definition for an atomic claim:

**Definition 1.** *An atomic claim delineates an indivisible minimal fact, which either describes a property of an individual or the relationship between individuals.*

An atomic claim should possess the following properties:

**Indivisibility.** An atomic claim should delineates a single, indivisible fact, which means that it cannot be decomposed into simpler, more fundamental claims without loss of meaning.

**Semantic Integrity.** An atomic claim should contain sufficient information to preclude any inconsistencies or ambiguities with the original claim.

**Verifiability.** An atomic claim should be verifiable. This means that it should delineates a fact that is worthy to be verified, rather than a commonsense or subjective opinion.

**Context Independence.** An atomic claim should be context-independent, therefore its truthfulness can be assessed independently without preceding and following context.

## 3.2   Atomic Claim Decomposition Task

This study primarily focuses on the Long-form Question Answering (LFQA) task. Long-form question answering aims at answering complex, open-ended questions with detailed, paragraph-length responses [14]. We define the task of atomic claim decomposition within this scenario. Given a question $q$ and a long response $r$ generated by LLMs, the goal is to decompose the response into a list of atomic claims $C = \{ac_1, ac_2, ..., ac_n\}$. Assuming the decomposition method is denoted by $D$, this can be formulated into the following function:

$$D(q, r) = \{ac_1, ac_2, ..., ac_n\} \tag{1}$$

It should be noted that only the facts within the response require decomposition. Hence, the decomposition method must be capable of identifying nonfactual sentences within the response and disregarding them.

**Fig. 2.** A example of the data annotation pipeline.

## 4   Chinese Atomic Claim Decomposition Dataset

To facilitate research on atomic claim decomposition task, we introduce Chinese Atomic Claim Decomposition Dataset, a dataset designed for claim decomposition task within the LFQA scenario. We first select data from WebCPM [14] dataset and then annotate them by our human-annotation pipeline, which includes three steps: sentence decomposition, sentence classification and atomic claim decomposition. Figure 2 shows a example of our data annotation pipeline.

## 4.1   Data Collection

WebCPM, an innovative open-source project, is designed to advance the field of interactive search research. Utilizing Chinese pre-trained models, this project imitates human web search behaviors and answers questions based on the facts collected from the websites it returns. It introduces a dataset that contain questions and LLM generated answers, which is suitable for our task. Therefore, We sample the first ten percent (550 of 5500) of WebCPM dataset to serve as the foundation for our subsequent manual annotation.

## 4.2   Data Annotation

Our data annotation pipeline can be divided into three parts. First, we decompose the whole response into sentences. Then, we classify each sentence into four categories. Finally, for sentences that have been labeled as fact, we decompose them into atomic claims.

**Sentence Decomposition.** Given a response generated by LLMs, it is infeasible to directly decompose it into atomic claims. Thus, we first decompose it into nature language sentences using Natural Language Toolkit (NLTK), which allows annotators to focus more on the decomposition of atomic claims within individual sentences, thereby reducing the difficulty of annotation. Along with sentence decomposition, we also filter out the citation marks in the responses.

**Sentence Classification.** Not all claims in a response are worthy to be verified. For example, subjective opinions like *"I think the iPad 10 has a very high cost performance ratio."* or instructions like *"Don't blindly follow the trend anymore."* do not need to check their truthfulness. A claim is considered checkworthy if it is one for which the general public has an interest in knowing the truth [8]. However, we assume that people who ask LLMs are interested in all factual claims in the response.

Considering the characteristics of LFQA data, we classify the sentences into four categories: fact, opinion, instruction and others. For each category, we provide a definition along with the subcategories it encompasses.

**Fact.** A fact is defined as a description of an objectively existing entity or event. Facts are objective, verifiable, and not influenced by personal beliefs or opinions. For example, "the main destructive power of nuclear bombs comes from the shock wave effect." is a fact.

**Opinion.** A opinion is defined as a belief or judgment that is not necessarily based on absolute certainty or proof. It is a personal perspective or view that may be influenced by subjective interpretations, emotions, or biases. For example, "Stephen Hawking is the greatest scientist in history." is a opinion.

**Instruction.** An instruction refers to a directive or command that specifies how a procedure or process should be executed. It outlines the actions to be taken to achieve a particular outcome or to operate a system or device. For example, "Don't blindly follow the trend anymore." is a instruction.

**Other.** This category contains sentences that cannot be classified into the three categories mentioned above, such as sentences that are rhetorical questions or connect the context.

**Atomic Claim Decomposition.** As mentioned above, we only perform atomic claim decomposition on sentences that are classified as facts in the previous step. Following the definition of atomic claims outlined in Sect. 3.1, factual sentences are decomposed into context-independent, verifiable atomic claims. During the decomposition process, annotators can simultaneously view the sentence to be decomposed along with the preceding and following sentences, as well as the questions, which provide both local and global context for the decomposition. This is beneficial for annotators to perform anaphora resolution.

### 4.3 Dataset Analysis

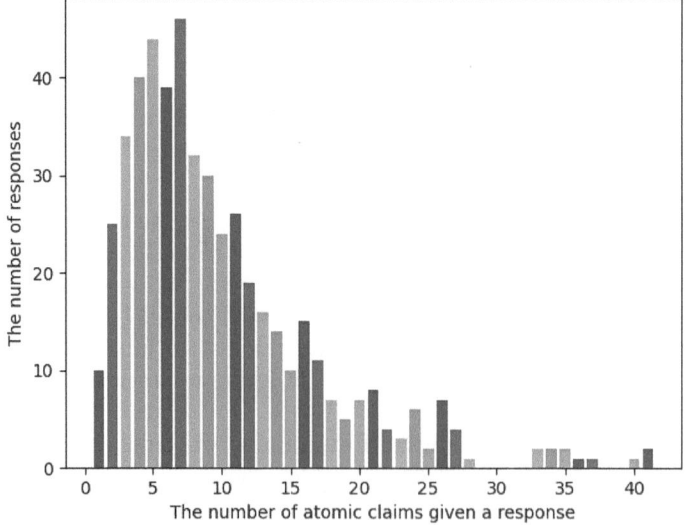

**Fig. 3.** The distribution of atomic claims amount given a response.

**Statistics.** We have retained 500 question-answer pair data containing a list of atomic declarations, totaling 4,956 atomic claims. The average atomic claims of each response is 9.912, which means that, on average, each response contains nearly ten distinct atomic claims. Figure 3 provides insight into the density of facts within the responses. It also suggests that the responses are rich in detail and that the decomposition process effectively captures the multiple atomic claims within each factual sentences.

**Sentences.** Most responses contain less than 10 sentences, while the longest response contain 39 sentences and there are 5 responses contain more than 21 sentences. Figure 4 shows the number of responses which contain less than 21 sentences. In addition, as shown in Fig. 5, most of the sentences are facts, totaling 2,484 of 2,907 sentences. 258 sentences are considered to be opinion and 64 are instruction.

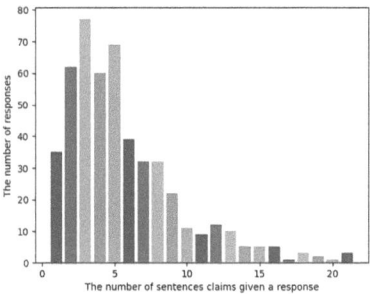 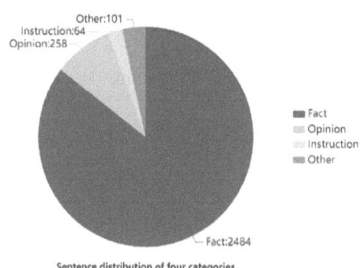

**Fig. 4.** The distribution of sentences amount given a response

**Fig. 5.** The distribution of sentences categories.

## 5   Experiments

### 5.1   Experimental Setup

We use CACDD dataset for atomic claim decomposition task with multi advanced LLMs. Open source LLMs include Baichuan2-7B, Glm4-9B, Llama3-8B, Mistral-7B, Qwen2-7B and Solar-10.7B. Moreover, we also measure the performance of GPT3.5-turbo. For better instruction compliance, we choose the chat or instruct version of the LLMs mentioned above. We prompt these LLMs under two settings: zero-shot and three-shot. For zero-shot, we just introduce the atomic claim definition to the model and instruct it to decompose the whole response in the prompt. For three-shot, we carefully selected three examples as context information for models to learn. In detail, we selected examples that require anaphora resolution or ellipsis supplementation from the context of the question or answer, and contain multiple factual and non factual sentences at the same time. All prompts can be found in our github repository. To facilitate reproducibility, all of our experiments set the "do_sample" parameters to false and employ greedy decoding. We also fine-tune a Llama3-8B model using 350 data from the CACDD dataset as strong baseline.

## 5.2   Metrics

For evaluation, we report the precision, recall and F1 scores of three widely-used metrics of text similarity: EM, Rouge-l and BertScore. For the Rouge-l metric, we set the threshold for text match to 0.8, while for the BertScore metric it is 0.9.

## 5.3   Results

**Table 1.** Zero-shot results of LLMs on atomic claim decomposition.

| Model | EM | | | Rouge-l | | | BertScore | | |
|---|---|---|---|---|---|---|---|---|---|
| | Precision | Recall | F1 | Precision | Recall | F1 | Precision | Recall | F1 |
| Baichaun2-7B | **9.28** | 5.74 | 6.25 | 29.68 | 18.35 | 20.30 | 29.87 | 18.53 | 20.56 |
| Glm4-9B | 4.89 | 6.00 | 5.21 | **31.57** | **39.70** | **33.30** | 30.68 | **38.55** | **32.38** |
| Llama3-8B | 0.17 | 0.44 | 0.23 | 16.21 | 27.41 | 18.98 | 15.06 | 25.16 | 17.56 |
| Mistral-7B | 2.64 | 2.79 | 2.53 | 18.75 | 18.78 | 17.75 | 20.47 | 20.51 | 19.37 |
| Qwen2-7B | 1.88 | 2.19 | 1.88 | 27.90 | 27.02 | 26.04 | 28.72 | 27.65 | 26.79 |
| Solar-10.7B | 0.02 | 0.03 | 0.02 | 0.50 | 0.77 | 0.56 | 0.61 | 0.92 | 0.68 |
| GPT3.5-turbo | 7.98 | **9.70** | **8.26** | 27.23 | 31.75 | 27.64 | 28.36 | 32.48 | 28.45 |

Table 1 and 2 show the results of our experiments. For zero-shot setting, Chinese open-source LLMs, including Glm4-9B, Baichuan2-7B and Qwen2-7B, show their better performance on this task. Glm4-9B achieves the best results on the Rouge-l and BertScore metrics, demonstrating its strong Chinese comprehension ability. The reason is that these models are enhanced for Chinese data, and can better understand and follow Chinese prompt. However, the performance of models that are not enhanced perform poorly on zero-shot setting. For example, the performance of Solar-10.7B is poor, as its responses contain a significant number of English sentences, as well as Chinese sentences that interspersed with Korean phrases. GPT3.5 achieves the best results on EM metric, which shows its excellent Chinese understanding and language ability. Nevertheless, the performance is not satisfactory yet.

For three-shot setting, Llama3-8B achieves the best results on all of the three metrics due to its strong capability in following instructions and learning from context. In addition, with carefully selected examples, Solar-10.7B show its excellent performance second only to Llama3-8B. This result, in contrast to the zero-shot setting, indicating that the model's ability of instruction following plays a much greater role than its fundamental language ability when presented with just a limited number of examples. However, all of these results are comparatively not satisfactory, indicating that there is significant room for improvement in the performance of LLMs on this task.

**Table 2.** Three-shot results of LLMs on atomic claim decomposition.

| Model | EM | | | Rouge-1 | | | BertScore | | |
|---|---|---|---|---|---|---|---|---|---|
| | Precision | Recall | F1 | Precision | Recall | F1 | Precision | Recall | F1 |
| Baichaun2-7B | 15.72 | 12.91 | 13.26 | 38.73 | 31.04 | 32.43 | 39.20 | 31.55 | 32.89 |
| Glm4-9B | 9.32 | 11.15 | 9.71 | 35.14 | 41.87 | 36.51 | 36.68 | 43.83 | 38.12 |
| Llama3-8B | **20.28** | **23.04** | **20.62** | **43.67** | **48.73** | **44.09** | **43.89** | **48.89** | **44.30** |
| Mistral-7B | 9.47 | 9.36 | 8.95 | 36.39 | 35.36 | 34.21 | 37.47 | 36.38 | 35.12 |
| Qwen2-7B | 11.60 | 12.59 | 11.41 | 37.99 | 39.16 | 36.57 | 38.96 | 40.07 | 37.43 |
| Solar-10.7B | 14.04 | 16.41 | 14.44 | 40.95 | 46.08 | 41.45 | 41.12 | 46.35 | 41.66 |
| GPT3.5-turbo | 11.74 | 11.58 | 11.09 | 39.05 | 36.06 | 35.51 | 39.81 | 36.76 | 36.13 |
| Llama3-8B-FT | **40.56** | **44.43** | **41.12** | **63.11** | **68.37** | **63.52** | **65.33** | **70.60** | **65.66** |

Fine-tuned model shows best performance on atomic claim decomposition task. We split the CACDD dataset into training and testing sets, fine tune the model with 350 pieces of data, and test it with 150 pieces of data. The results indicate that fine-tuning with a small amount of high-quality data can significantly enhance the performance of LLMs on this task.

# 6   Conclusion

To tackle the limitation that previous research overlooks the importance of claim decomposition, we introduce a new Chinese dataset for the atomic claim decomposition task. This is the first Chinese dataset designed for this task under the real-world LFQA scenario. Following Bertrand Russell's philosophy of logical atomism, we provide a definition of atomic claims and a pipeline to annotate atomic claim data. Our experimental results show that existing LLMs still have significant room for improvement on this task. In the future, we intend to build a larger-scale dataset and propose a new approach to enhance the capacity of LLMs for the atomic claim decomposition task.

**Limitations.** The CACDD dataset only contains 500 question-answer pairs, which is too small to comprehensively evaluate the LLMs. Furthermore, we assume that all facts in the LLM generated responses are of interest to the user, but some people may have different views on this assumption.

**Acknowledgement.** This work was funded by the National Natural Science Foundation of China (NSFC) under Grants No. 62372431 and 62472408, the Strategic Priority Research Program of the CAS under Grants No. XDB0680102, XDB0680301, the National Key Research and Development Program of China under Grants No. 2023YFA1011602, the Youth Innovation Promotion Association CAS under Grants No. 2021100, the Lenovo-CAS Joint Lab Youth Scientist Project, and the project under Grants No. JCKY2022130C039.

# References

1. Chen, J., Kim, G., Sriram, A., Durrett, G., Choi, E.: Complex claim verification with evidence retrieved in the wild. In: Proceedings of the 2024 Conference of the North American Chapter of the Association for Computational Linguistics: Human Language Technologies (Volume 1: Long Papers), pp. 3569–3587 (2024)
2. Chen, J., Sriram, A., Choi, E., Durrett, G.: Generating literal and implied subquestions to fact-check complex claims. In: Proceedings of the Conference on Empirical Methods in Natural Language Processing (EMNLP) (2022)
3. Chen, S., et al.: Sub-sentence encoder: contrastive learning of propositional semantic representations. In: Proceedings of the 2024 Conference of the North American Chapter of the Association for Computational Linguistics: Human Language Technologies (Volume 1: Long Papers), pp. 1596–1609 (2024)
4. De Marneffe, M.C., Manning, C.D., Nivre, J., Zeman, D.: Universal dependencies. Comput. Linguist. **47**(2), 255–308 (2021)
5. Gao, L., et al.: RARR: researching and revising what language models say, using language models. In: Proceedings of the 61st Annual Meeting of the Association for Computational Linguistics (Volume 1: Long Papers), pp. 16477–16508 (2023)
6. Gao, T., Yen, H., Yu, J., Chen, D.: Enabling large language models to generate text with citations. In: Proceedings of the 2023 Conference on Empirical Methods in Natural Language Processing, pp. 6465–6488 (2023)
7. Gao, Y., Sun, C., Passonneau, R.J.: Automated pyramid summarization evaluation. In: Proceedings of the 23rd Conference on Computational Natural Language Learning (CoNLL) (2019)
8. Hassan, N., Li, C., Tremayne, M.: Detecting check-worthy factual claims in presidential debates. In: Proceedings of the 24th ACM International on Conference on Information and Knowledge Management, pp. 1835–1838 (2015)
9. Jing, L., Li, R., Chen, Y., Jia, M., Du, X.: Faithscore: evaluating hallucinations in large vision-language models. arXiv preprint arXiv:2311.01477 (2023)
10. Kamoi, R., Goyal, T., Rodriguez, J.D., Durrett, G.: Wice: real-world entailment for claims in Wikipedia. In: Proceedings of the 2023 Conference on Empirical Methods in Natural Language Processing, pp. 7561–7583 (2023)
11. Lemmon, E.J.: Comments on d. davidson's "the logical form of action sentences"'. In: The Logic of Decision and Action, pp. 96–103 (1967)
12. Min, S., et al.: Factscore: fine-grained atomic evaluation of factual precision in long form text generation. In: Proceedings of the 2023 Conference on Empirical Methods in Natural Language Processing, pp. 12076–12100 (2023)
13. Niklaus, C., Cetto, M., Freitas, A., Handschuh, S.: Dissim: a discourse-aware syntactic text simplification framework for English and German. In: Proceedings of the 12th International Conference on Natural Language Generation, pp. 504–507 (2019)
14. Qin, Y., et al.: WebCPM: interactive web search for Chinese long-form question answering. In: Proceedings of the 61st Annual Meeting of the Association for Computational Linguistics (Volume 1: Long Papers), pp. 8968–8988 (2023)
15. Russell, B.: The philosophy of logical atomism: lectures 1–2. Monist **28**(4), 495–527 (1918)
16. Shi, W., Han, X., Lewis, M., Tsvetkov, Y., Zettlemoyer, L., Yih, W.t.: Trusting yourevidence: hallucinate less with context-aware decoding. In: Proceedings of the 2024 Conference of the North American Chapter of the Association for Computational Linguistics: Human Language Technologies (Volume 2: Short Papers), pp. 783–791 (2024)

17. Wang, X., et al.: Self-consistency improves chain of thought reasoning in language models. In: The Eleventh International Conference on Learning Representations (2023)
18. Wang, Y., et al.: Factcheck-GPT: end-to-end fine-grained document-level fact-checking and correction of LLM output. arXiv preprint arXiv:2311.09000 (2023)
19. Wanner, M., Ebner, S., Jiang, Z., Dredze, M., Van Durme, B.: A closer look at claim decomposition. arXiv preprint arXiv:2403.11903 (2024)
20. White, A.S., et al.: Universal decompositional semantics on universal dependencies. In: Proceedings of the 2016 Conference on Empirical Methods in Natural Language Processing, pp. 1713–1723 (2016)
21. Yue, X., Wang, B., Chen, Z., Zhang, K., Su, Y., Sun, H.: Automatic evaluation of attribution by large language models. In: Findings of the Association for Computational Linguistics: EMNLP 2023, pp. 4615–4635 (2023)

# Dual-Granularity Hierarchical Fusion Network for Multimodal Humor Recognition on Memes

Mengyi Wang[1], Shuo Hou[1], Hongfei Lin[2], and Yijia Zhang[1]($\boxtimes$)

[1] School of Information Science and Technology, Dalian Maritime University,
Dalian 116024, China
{mengyiw,hs3020664}@dlmu.edu.cn, hflin@dlut.edu.cn
[2] School of Computer Science and Technology, Dalian University of Technology,
Dalian 116024, China
zhangyijia@dlmu.edu.cn

**Abstract.** Multimodal humor recognition has gradually drawn attention in recent years, with memes serving as a prominent form of multimodal communication on the internet. However, current research in multimodal humor recognition lags behind due to the lack of publicly available datasets. To address this gap, we create the HuME meme dataset, consisting of 10,600 images paired with Chinese text, aimed at discerning whether memes are humorous. Furthermore, given humour's metaphorical nature, existing multimodal models struggle to fully adapt to humor recognition tasks. Hence, we delve into humor theory, analyzing the textual and visual coherence within memes from syntactic and semantic perspectives. Finally, considering the high-dimensional nature of the data, we incorporate manifold learning to further represent the features of high-dimensional data. We quantitatively and qualitatively analyze the experimental results to verify the legitimacy of the HuME dataset and the effectiveness of **D**ual-granularity **H**ierarchical **F**usion **N**etwork. The dataset is available at https://github.com/DericWmy/HuME.

**Keywords:** Multimodal humor recognition · Humorous meme · Manifold learning

## 1 Introduction

Humor is a complex psychological and social phenomenon that involves people's ability to understand, express, or explain things in some creative or unexpected way [4]. Furthermore, humor assessment often relies on contextual cues and other non-linguistic information, attempting to subvert the reader's common sense or preconceptions. Humor recognition is a very challenging research topic in natural language processing. Memes are considered widespread Internet cultural genes that spread in a derivative way, spreading in the form of various cartoon images that people like to see and hear [12]. They can express various emotions, such as humor, sarcasm, malice, etc. However, such speech is usually highly

X. He et al. (Eds.): CCIR 2024, LNCS 15418, pp. 54–65, 2025.
https://doi.org/10.1007/978-981-96-1710-4_5

metaphorical, and its true intention often differs from the literal meaning. The expressed intention is the opposite. Memes often contain implicit humor, which can mislead the judgment of humor by multimodal models. For example, it can use non-humorous images and humorous text, or vice versa. In Fig. 1, we give an example of a meme that uses images as clues to compare a lynx and a kitten to a teacher and a student, respectively, mapping humor in a contrasting way.

**Fig. 1.** The original memes(right) and their non-humor version(left) (Meme texts in every figure has been translated to English for better demonstration.)

Currently, the application of multimodal models in detecting the humor in memes faces several challenges [22]. Firstly, within the multimodal domain, there is a notable absence of high-quality related datasets. Secondly, the textual content encompassed within memes varies significantly, ranging from single words to short phrases or incomplete sentences. As a result, it becomes challenging to extract complete syntactic information from the text and complicates the utilization of mainstream multimodal models for the detection of humor.

We propose to construct a diverse meme dataset to address the current challenges arising from the lack of variety in humor meme datasets. Simultaneously, we propose a multimodal model that takes a hierarchical approach to progressively address the semantic consistency between text modality and image modality. In the coarse-grained cross-modal consistency module, the model employs a multi-head cross attention mechanism to align images and text while further calculating their semantic consistency. In the fine-grained cross-modal consistency module, an extension of the previous level, we investigate the spatial and semantic relationships between images and text. We attempt to use the concept of manifold learning for modeling intra-modality relationships to aid in obtaining semantic consistency relationships between image modality and text modality. The primary contributions of our work are outlined as follows

- We construct a dataset comprising 10,600 Chinese memes, each of which has been manually labeled with humor or non-humor tags.
- We propose a hierarchical fusion network to explore coarse-grained and fine-grained consistent semantic relationships between different modalities, and

incorporate a locally linear embedding method to better capture the intrinsic nonlinear structure of the data.
- We select multiple baseline models to validate our model and dataset, further corroborating the effectiveness of our model and the feasibility of our dataset.

## 2   Related Work

### 2.1   Text-Based Humor Recognition

Early works typically treated humor recognition as a text classification task without a deeper understanding of the mechanisms of humor. Fan et al. [7] integrated additional phonetic and semantic(ambiguity) information into a deep learning framework. Xie et al. [20] established connections between the GPT-2 pre-trained model and humor theories, proposing and evaluating humor uncertainty and surprise. Xie et al. [21] employed the disentangled attention mechanism of the DeBERTa architecture [10] for humor label prediction. Ren et al. [16] approached humor recognition from a perspective of humor linguistics and developed a neural network named ANPLS. This framework utilized speech understanding units to extract humor features, vocabulary understanding units to resolve lexical ambiguities, and context understanding units to capture contextual humor features, thus addressing inconsistencies and fuzziness in humor. In addition, Zhai et al. [23] proposed a generative network to improve the performance of generating emotionally rich and humorous responses while taking cultural diversity into account.

### 2.2   Multimodal Humor Recognition

Relying solely on textual information for humor recognition tasks fails to capture non-verbal information from modalities such as images and audio, thus limiting the understanding of humorous expressions. Multimodal humor recognition is a task that encompasses multiple data modalities, including text, images, and audio. Hasan et al. [8] introduced the Humor Knowledge-rich Transformer(HKT) to integrate preceding context and external knowledge, capturing critical elements in multimodal humor detection. The MuLOT proposed by Pramanick et al. [15] effectively captures dynamic intra- and inter-modal dependencies highly reliant on various modalities, even when training data is limited. Bedi et al. [1] proposed MSH-COMICS, an attention-based multimodal classification model to learn rich discourse textual representations.

In a study by Hasan et al. [9], a UR-FUNNY dataset was introduced, comprising three modalities: text, speech, and images, to analyze humorous emotions. The dataset extracted 16,514 pieces of speech modality data from TED speeches, with each speech segment composed of dialogues. In addition, Wu et al. [19] presented the MUMOR humor recognition corpus, which contains Chinese dialogues derived from 1,298 dialogues extracted from two TV sitcoms. Bertero et al. [2] constructed a dataset consisting of 43,672 lines of dialogue by collecting letters and scripts from "The Big Bang Theory." Their research combines text and sound modalities to detect humor in conversations.

## 2.3 Manifold Learning

In the original high-dimensional space, there exists redundant and noisy information, which introduces errors in practical applications, thereby affecting accuracy. Since Principal Component Analysis(PCA) needs to consider the distribution and characteristics of data samples in the high-dimensional space, it fails to capture the complex structures present in the data. Jiang et al. [11] proposed that manifold learning is used to capture and utilize the intrinsic manifold structure to calculate more effective distance and similarity measures, aiming to reveal intrinsic structures and patterns in high-dimensional data. Pedronette et al. [14] made relevant progress in context similarity measurement using unsupervised manifold learning methods based on ranking information. Valem et al. [18] proposed a semi-supervised framework based on manifold learning and GCN model for image classification tasks.

# 3 Dataset

## 3.1 Data Collection and Annotation

We use crawler programs to collect the data required for our experiments from the current mainstream communities. The primary sources of data are communities such as Weibo and Twitter. In the initial data acquisition stage, we use #meme and #memes as our search keywords. We use Baidu AI and iFlytek API to perform Optical Character Recognition(OCR) on the obtained images to obtain the captions in memes. Since the humor recognition task itself is highly subjective, it is easy to cause negative impact on the annotator during the data annotation process. Therefore, we first screen the professional background of the data annotators to ensure that the gender, age, race, region and research direction of the data annotators are as wide as possible, and ensure that they have rich field experience. We have counted the statistical data of the annotators as shown in Table 1.

**Table 1.** Annotators demographics.

| Characteristic | Demographics |
| --- | --- |
| Gender | 4 male, 4 female |
| Age | 8 age < 25, 7 age $\leq$ 25 |
| Race | 6 Asian, 2 others |
| Region | From 5 different provinces |
| Education | 1 BD, 4 MA, 3Ph.D. |

We use Fleiss' Kappa [3] score as a measure of consistency between annotators. After the pre-annotation stage, the final Fleiss' Kappa score finally reached

0.772, indicating that the consistency between annotators has met the standard. After achieving a reasonable consensus among all annotators, we proceed to the formal annotation phase. We distribute all memes to annotators, instructing them to perform the labeling task strictly according to the prescribed guidelines. The majority vote is then employed to determine the gold label, which is subsequently integrated into the entire dataset. We visualized the entire process of data collection and annotation in the format of a flowchart, as illustrated in Fig. 2.

**Fig. 2.** The flowchart of the collection and annotation process.

After completing the entire data annotation process, we divided the experimental data into training, validation, and test sets in a ratio of 80%:10%:10%, ensuring a consistent distribution of positive and negative samples. This partitioning was done to maintain the integrity and fairness of the dataset. Our dataset statistics are shown in Table 2.

**Table 2.** The statistics of our dataset.

| Dataset | Train | Valid | Test | Total |
|---|---|---|---|---|
| Humorous | 4,831 | 604 | 604 | 6,039 |
| Non Humorous | 3,660 | 451 | 450 | 4,561 |

## 4   Methodology

In this section, we will introduce the baselines and methods for multi-modal humor recognition using our dataset. DHFN is mainly divided into three parts, namely feature extraction module, coarse-grained cross-modal consistency module(CCCM), and fine-grained cross-modal consistency module(FCCM). The overall framework of DHFN is shown in Fig. 3.

### 4.1   Feature Extraction Module

In the feature extraction module, given an input text-image pair $\{X_T, X_I\}$, the corresponding text features and image features are generated through the text encoder and image encoder respectively.

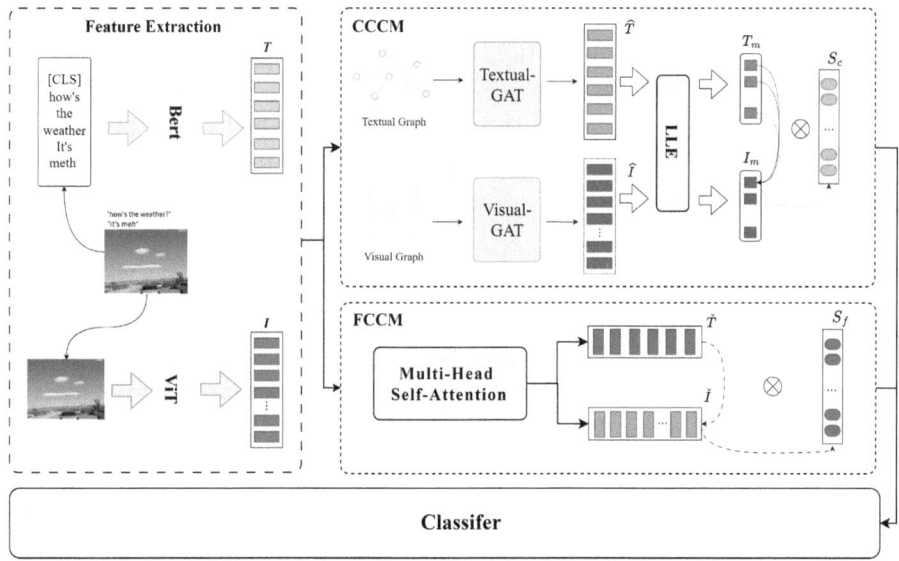

**Fig. 3.** The overall framework of DHFN.

**Text Encoder.** We use the pre-trained model BERT [5] as our text encoder to generate text features. Specifically, we input the text $X_T = \{w_1, w_2, \ldots, w_n\}$ composed of $n$ tokens into BERT to obtain the text feature $T$, which is expressed as $T = Bert(X_T) = [t_1, t_2, \ldots, t_n]$ where $T \in \mathbb{R}^{n \times d}$.

**Image Encoder.** Vision Transformer (ViT) [6] is pretrained on large-scale image datasets such as ImageNet, and introduces a self-attention mechanism in the visual field. Therefore we use the ViT pre-trained model to generate image features. In the image preprocessing stage, each image is first resized to 224×224, and each image is divided into $p$ image patches of equal size. Finally, the generated image patches are reconstructed into a flat patch sequence. Input the processed image patch sequence into ViT to obtain the image features, expressed as $I = ViT(X_I) = [v_1, v_2, \ldots, v_p]$ where $I \in \mathbb{R}^{p \times d}$.

### 4.2 Coarse-Grained Cross-Modal Consistency Module

This module uses multihead cross-attention to calculate the semantic consistency relationship between image and text sequences. We employ a cross-attention mechanism with $n$ heads to map the text modality and the image modality to the same semantic space, which calculation process is as follows

$$head_i = softmax(\frac{Q_i^T}{\sqrt{d/h}}K_i^I)V_i^I \tag{1}$$

We follow the method of [13] and use text as query and image as key and value, where $Q_i^T = W_i^q T$, $K_i^I = W_i^k I$ and $V_i^I = W_i^v I$, specifically, $W_i^q \in \mathbb{R}^{d \times \frac{d}{h}}$, $W_i^k \in \mathbb{R}^{d \times \frac{d}{h}}$ and $W_i^v \in \mathbb{R}^{d \times \frac{d}{h}}$ are parameter matrices.

Then all heads are connected through two layers of multi-layer perceptrons and residual connections, and an updated text feature representation is obtained after a regularization layer as

$$\check{T} = LN(MLP(contact\,[head_i]_{i=1}^{h} + T)) \tag{2}$$

where LN denotes the layernorm regularization layer, MLP denotes the multi-layer perceptron, and $contact\,(\cdot)$ denotes the concatenation operation.

A set of consistency scores are calculated through the softmax function, and we use the consistency scores to measure the importance of each word for the humor recognition task. The consistency score between the patch and the text token is expressed as

$$s_c = softmax(\check{T}W_c + b_c)^{\top} Q_c \tag{3}$$

where $W_c \in \mathbb{R}^{d \times 1}$ and $b \in \mathbb{R}^n$ are parameters that can be learned. We use inner products to detect consistency between modalities, specifically $Q_c = \frac{(\check{T}I^{\top})}{\sqrt{d}}$, where $Q_c \in \mathbb{R}^{n \times p}$. Specifically, $Q_c[i, j]$ represents the similarity score between the $i$-th token and the $j$-th patch.

## 4.3   Fine-Grained Cross-Modal Consistency Module

Multimodal humor recognition often relies on contextual information, so the premise of understanding memetic humor needs to be based on understanding textual information. It is particularly important to obtain syntactic information of the text. Spacy uses dependency syntactic analysis to identify the dependencies between words, so we use it to build the corresponding text dependency graph. Specifically, if there is a dependency between two words after word segmentation, it is reflected in the corresponding edge in the text graph to represent their relationship. The complete text dependency graph is built through continuous iteration. Similarly, we construct a visual map based on the relative positions between image patches. We then input the graphs of text modality and visual modality into the GAT for further processing.

In this module, we take the constructed text dependency graph as an example, where $\theta_l \in \mathbb{R}^{2d}$ and $W_l \in \mathbb{R}^{d \times d}$ respectively represent the parameters that can be learned in the $L$-th layer GAT, $\alpha_{i,j}^l$ represents the attention score between the $i$-th graph node and its neighbor node $j$ in the dependency graph, which can be expressed as

$$\alpha_{i,j}^l = \frac{exp(ReLU(\theta_l^{\top}[W_l t_i^l \parallel W_l t_j^l]))}{\sum_k exp(ReLU(\theta_l^{\top}[W_l t_i^l \parallel W_l t_k^l]))} \tag{4}$$

$$t_i^{l+1} = \theta_l \alpha_{i,i}^l t_i^l + \sum_{j \in N(i)} \theta_l \alpha_{i,j}^l t_i^l \tag{5}$$

We define the initialized text feature $t_0$, and the text feature input generated by each layer is output by the GAT feature of the previous layer, where $N(i)$ represents all neighbors of the $i$-th node, $t_i$ represents the characteristic representation of the $i$-th node, and $t_j$ represents the characteristic representation of its neighbor node $j$. After that, we get the final embedding $\hat{T} = [t_1^L, t_2^L, \ldots, t_n^L]$. Similarly, we take the positional map of the image modality as input and employ the same method to obtain the updated representation $\hat{I} = [i_1^L, i_2^L, \ldots, i_n^L]$ of the image.

**Manifold Learning.** The locally linear embedding method achieves data dimensionality reduction by minimizing the difference between the similarity between high-dimensional data points and the similarity between low dimensional data points. Specifically, we use the local linear embedding algorithm as our representative method to explore the nonlinear structure of high-dimensional features. And we define the manifold learning method as the function $LLE_m$. The function $LLE_m$ takes the latent vector generated by GAT as input and returns a set of vectors after nonlinear dimensionality reduction as output Tm, where the original high-dimensional data is mapped to low-dimensional features by the unsupervised local linear embedding algorithm while retaining the original information as much as possible. The formula can be expressed as

$$T_m = LLE_m(\hat{T}) \tag{6}$$

At last we calculate the semantic consistency score $s_f$ as

$$s_f = softmax(T_m W_f + b_f)^\top Q_f \tag{7}$$

where $W_c \in \mathbb{R}^{d \times 1}$ and $b \in \mathbb{R}^n$ are parameters that can be learned. Similarly, we use the inner product $Q_f$ to represent the consistency between text and image modalities at the enhancement level, where $Q_f = \frac{(\hat{T}\hat{I}^\top)}{\sqrt{d}}$, specifically, $Q_f \in \mathbb{R}^{n \times p}$, $\hat{T} \in \mathbb{R}^{n \times d}$, $\hat{I} \in \mathbb{R}^{p \times d}$.

## 4.4 Classification

Through the consistency scores $s_c$ and $s_f$ obtained by the Coarse-grained cross-modal consistency module and the Fine-grained cross-modal consistency module, we obtain the final prediction result of the label as

$$\hat{y} = softmax(W_y[p_y \odot s_c \parallel p_y \odot s_f] + b_y) \tag{8}$$

$$p_y = softmax(W_v I + b_v) \tag{9}$$

where $W_y \in \mathbb{R}^{2 \times 2p}$, $W_v \in \mathbb{R}^{d \times 1}$, $b_y \in \mathbb{R}^2$, $b_v \in \mathbb{R}^p$ are parameters that can be learned, and $\odot$ denotes element-wise vector product.

Finally, we employed the commonly used cross-entropy loss in classification tasks to optimize our model as follows

$$\mathcal{L} = -\sum_{i=0}^{N} y_i^{\top} log \hat{y}_i \tag{10}$$

where $y$ is the true label and $\hat{y}$ represents the probability of the predicted label of the $i$-th text-image pair.

## 5   Experiments and Discussion

### 5.1   Experimental Results

Based on the dataset proposed in this paper, the performance comparison results of DHFN and other baseline models are shown in Table 3. Our analysis reveals that textual content inherently contains richer information compared to images. Multimodal models can extract relevant information from text and images and thus perform better than unimodal model. Specifically, we compare DHFN with the existing state-of-the-art baseline model DynRT-Net [17] and find that DHFN achieves 1.84% higher accuracy and 1.11% higher F1 score. Our proposed multimodal approach demonstrates significantly better performance than previous methodologies, thus validating the effectiveness of DHFN in integrating features from different modalities at multiple levels.

**Table 3.** The performance comparison results between DHFN and the baseline model on our dataset are presented (Abbreviation: Acc/P/R/F1: Accurary/Precision/Recall/F1-score). The best results are highlighted by boldface font.

| Modality | Model | Acc(%) | P(%) | R(%) | F1(%) |
|---|---|---|---|---|---|
| Text-Only | BERT | 68.49 | 67.42 | 67.86 | 67.61 |
| Image-Only | ViT | 66.28 | 67.38 | 67.57 | 66.26 |
| | ResNet | 66.02 | 65.48 | 65.29 | 65.37 |
| Multimodal | ViLBERT CC | 71.53 | 70.82 | 70.32 | 70.29 |
| | ViLBERT COCO | 71.88 | 71.21 | 70.84 | 71.05 |
| | MOMENTA | 72.37 | 72.03 | 71.55 | 71.72 |
| | DynRT-Net | 73.42 | 73.01 | 72.11 | 72.43 |
| | **DHFN** | **75.26** | **75.64** | **73.05** | **73.54** |

### 5.2   Ablation Study

To further demonstrate the impact of different components of DHFN on the overall performance, we conduct the following ablation experiments and list the results of our study in Table 4.

We first removed the coarse-grained cross-modal consistency module during the experiment, and the accuracy and F1 score dropped by 0.96% and 0.98%

**Table 4.** The comparison results between DHFN and baseline models on our dataset.

| Model | Acc(%) | F1(%) |
|---|---|---|
| **DHFN** | **75.26** | **73.54** |
| w/o CCCM | 74.30 | 72.56 |
| w/o FCCM | 72.66 | 71.85 |
| w/o LLE | 73.70 | 72.99 |

respectively. Demonstrate that the model can detect inconsistencies between image patches and tokens. After removing the fine-grained cross-modal consistency module, the model performance dropped significantly compared to other modules, with the accuracy dropping by 2.6% and the F1 score dropping by 1.69%. Therefore, it is proved that after adding the combination of semantic dependence information and image position information, the model can be more conducive to discovering semantic inconsistencies between text and image information. At the same time, through the interaction between different modalities, it can reduce the information gaps between modalities. Finally we removed the module of data dimensionality reduction using LLE. After experimental comparison, we found that the accuracy of the experimental results, accuracy and F1 score decreased by 1.56% and 0.55% respectively. After analysis, we have reason to prove that by using the local linear embedding method to reduce the dimensionality of the data, most of the original information can still be maintained between the data to ensure the integrity and validity of the data.

### 5.3 Visualizations

We visualize the attention weight distribution of the model on images to generate visual explanations, as shown in Fig. 4. This visualization serves to validate the effectiveness of our proposed model and, simultaneously, to gain a deeper understanding of its interpretability. The image depicts a sign standing on the lawn with the message "Please Keep Off the Grass", creating a humorous sentiment in conjunction with the statement "I was just about to play football there". Through

**Fig. 4.** Visualization of attention weight distribution of DHFN.

the visualized attention weight map, where brighter colors indicate higher attention weights, it is evident that the attention is predominantly focused on the sign with the message "Please Keep Off the Grass". This observation underscores the efficacy of DHFN.

## 6    Conclusion and Future Work

In this study, we construct a dataset comprising 10,600 memes from various Chinese communities, which is used for the task of multimodal humor recognition in memes. We propose **D**ual-granularity **H**ierarchical **F**usion **N**etwork that integrates manifold learning methods to explore the consistency between text features and visual features in parallel, from both local semantics and global grammar. This approach enhances the accuracy of multimodal humor recognition. Rigorous experimental evaluations further validate the rationale of our proposed dataset and the superiority of the DHFN. We are committed to further exploring richer multimodal datasets and incorporating external knowledge to address the issue of insufficient visual feature knowledge.

**Acknowledgments.** This work is supported by grant from the Natural Science Foundation of China (No. 62076046).

## References

1. Bedi, M., Kumar, S., Akhtar, M.S., Chakraborty, T.: Multi-modal sarcasm detection and humor classification in code-mixed conversations. IEEE Trans. Affect. Comput. **14**(2), 1363–1375 (2023). https://doi.org/10.1109/TAFFC.2021.3083522
2. Bertero, D., Fung, P.: A long short-term memory framework for predicting humor in dialogues. In: Proceedings of the 2016 Conference of the North American Chapter of the Association for Computational Linguistics: Human Language Technologies, pp. 130–135 (2016)
3. Bobicev, V., Sokolova, M.: Inter-annotator agreement in sentiment analysis: machine learning perspective. In: International Conference Recent Advances in Natural Language Processing, pp. 97–102 (2017)
4. Chen, P.Y., Soo, V.W.: Humor recognition using deep learning. In: Proceedings of the 2018 Conference of the North American Chapter of the Association for Computational Linguistics: Human Language Technologies, vol. 2 (Short Papers), pp. 113–117 (2018)
5. Devlin, J., Chang, M.W., Lee, K., Toutanova, K.: BERT: pre-training of deep bidirectional transformers for language understanding. arXiv preprint arXiv:1810.04805 (2018)
6. Dosovitskiy, A., et al.: An image is worth $16 \times 16$ words: transformers for image recognition at scale. arXiv preprint arXiv:2010.11929 (2020)
7. Fan, X., et al.: Phonetics and ambiguity comprehension gated attention network for humor recognition. Complexity **2020**, 1–9 (2020)
8. Hasan, M.K., et al.: Humor knowledge enriched transformer for understanding multimodal humor. In: Proceedings of the AAAI Conference on Artificial Intelligence, vol. 35, pp. 12972–12980 (2021)

9. Hasan, M.K., Rahman, W., Zadeh, A., Zhong, J., Tanveer, M.I., Morency, L.P., et al.: UR-FUNNY: a multimodal language dataset for understanding humor. arXiv preprint arXiv:1904.06618 (2019)
10. He, P., Liu, X., Gao, J., Chen, W.: DeBERTa: decoding-enhanced BERT with disentangled attention. arXiv preprint arXiv:2006.03654 (2020)
11. Jiang, J., Wang, B., Tu, Z.: Unsupervised metric learning by self-smoothing operator. In: 2011 International Conference on Computer Vision, pp. 794-801. IEEE (2011)
12. Kiela, D., et al.: The hateful memes challenge: detecting hate speech in multimodal memes. Adv. Neural. Inf. Process. Syst. **33**, 2611–2624 (2020)
13. Liu, H., Wang, W., Li, H.: Towards multi-modal sarcasm detection via hierarchical congruity modeling with knowledge enhancement. arXiv preprint arXiv:2210.03501 (2022)
14. Pedronette, D.C.G., Valem, L.P., Almeida, J., Torres, R.D.S.: Multimedia retrieval through unsupervised hypergraph-based manifold ranking. IEEE Trans. Image Process. **28**(12), 5824–5838 (2019)
15. Pramanick, S., Roy, A., Patel, V.M.: Multimodal learning using optimal transport for sarcasm and humor detection. In: Proceedings of the IEEE/CVF Winter Conference on Applications of Computer Vision, pp. 3930–3940 (2022)
16. Ren, L., Xu, B., Lin, H., Zhang, J., Yang, L.: An attention network via pronunciation, lexicon and syntax for humor recognition. Appl. Intell. **52**(3), 2690–2702 (2022)
17. Tian, Y., Xu, N., Zhang, R., Mao, W.: Dynamic routing transformer network for multimodal sarcasm detection. In: Proceedings of the 61st Annual Meeting of the Association for Computational Linguistics (Volume 1: Long Papers), pp. 2468–2480 (2023)
18. Valem, L.P., Pedronette, D.C.G., Latecki, L.J.: Graph convolutional networks based on manifold learning for semi-supervised image classification. Comput. Vis. Image Underst. **227**, 103618 (2023)
19. Wu, J., Lin, H., Yang, L., Xu, B.: MUMOR: a multimodal dataset for humor detection in conversations. In: Natural Language Processing and Chinese Computing: 10th CCF International Conference, NLPCC 2021, Qingdao, China, October 13–17, 2021, Proceedings, Part I 10, pp. 619–627. Springer (2021). https://doi.org/10.1007/978-3-030-88480-2_49
20. Xie, Y., Li, J., Pu, P.: Uncertainty and surprisal jointly deliver the punchline: exploiting incongruity-based features for humor recognition. arXiv preprint arXiv:2012.12007 (2020)
21. Xie, Y., Li, J., Pu, P.: Humorhunter at semeval-2021 task 7: humor and offense recognition with disentangled attention. In: Proceedings of the 15th International Workshop on Semantic Evaluation (SemEval-2021), pp. 275–280 (2021)
22. Xu, H., et al.: Hybrid multimodal fusion for humor detection. In: Proceedings of the 3rd International on Multimodal Sentiment Analysis Workshop and Challenge, pp. 15–21 (2022)
23. Zhai, C., Wibowo, S.: A WGAN-based dialogue system for embedding humor, empathy, and cultural aspects in education. IEEE Access (2023)

# Exploring the Potential of Dimension Reduction in Building Efficient Dense Retrieval Systems

Zhipeng Xu[ID], Zhenghao Liu[(✉)][ID], Yu Gu[ID], and Ge Yu[ID]

Department of Computer Science and Technology, Northeastern University, Shenyang, China
xuzhipeng@stumail.neu.edu.cn,
{liuzhenghao,guyu,yuge}@mail.neu.edu.cn

**Abstract.** Dense retrievers utilize pretrained language models to encode queries and documents as high-dimensional embeddings for retrieval. Nevertheless, these high-dimensional embeddings usually result in more expensive index storage and higher retrieval latency. In this paper, we further explore the potential of building a lightweight dense retrieval system by combining the dimension reduction in the encoding-indexing pipeline. Our experiments demonstrate that the *encoding-compression-indexing-retrieval* method can conduct an efficient dense retrieval system, reducing the retrieval latency of 96%, while maintaining comparable retrieval effectiveness. Our further analyses illustrate that the dimensional reduction method can broaden its retrieval effectiveness in different domains and cooperate with different index building methods.

**Keywords:** Dense Retrieval · Dimension Reduction · Conditional Autoencoder

## 1 Introduction

Dense retrieval is commonly used as the first stage in several multi-stage NLP pipelines and has illustrated its effectiveness in searching external knowledge for various NLP tasks, such as question answering [6,9], fact verification [14,16], and conversational search [36]. Dense retrieval models aim to map both queries and documents into an embedding space and conduct semantic matching in such an embedding space to retrieve candidate documents that satisfy the information needs of the given queries [39]. Existing dense retrieval systems usually employ an *encoding-indexing-retrieval* pipeline to conduct efficient retrieval, which utilizes pretrained language models to encode both queries and documents and then uses some toolkits, such as faiss [8], to build the document index for retrieval.

The effectiveness of existing dense retrieval systems [2–4,9,25,29,32] mainly lies in the query and document encoding modules. Thus existing dense retrieval models keep the high-dimensional embeddings from pretrained language models to learn the relevance signals [18,26], *e.g.* 768-dimensional or 1024-dimensional embeddings for BERT (base) or BERT (large) model, respectively. These high-dimensional embeddings of dense retrievers are necessary to fit training signals and guarantee retrieval effectiveness during training [18,26]. Nevertheless, high dimensional embeddings usually exhaust the memory to store the index, leading to longer retrieval latency [5,22].

ⓒ The Author(s), under exclusive license to Springer Nature Singapore Pte Ltd. 2025
X. He et al. (Eds.): CCIR 2024, LNCS 15418, pp. 66–79, 2025.
https://doi.org/10.1007/978-981-96-1710-4_6

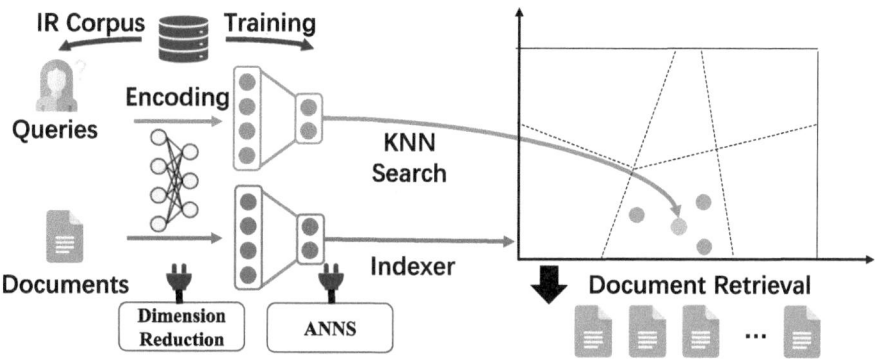

**Fig. 1.** The Architectures of Our **DI**mension **RE**du**CT**ion based ind**E**x buil**D**ing (DIRECTED) Framework.

For instance, NQ [10]–an open domain question answering dataset–contains over 21 million passages and requires approximately 61GB storage space for its flat document index alone. The sizeable document index frequently poses a hindrance in deploying dense retrieval or question-answering systems to different terminals and downstream applications. To conduct efficient dense retrieval, some studies focused on reducing the dimensions of document embeddings [15,18,35], which have yielded competitive result in vanilla dense retrieval. However, the effectiveness and generalization ability of dimension reduction methods in different retrieval scenarios still need to be further explored. We do not yet know how the dimension reduction approach can be combined with efficient index building or how it affects out-of-domain performance. For creating a self-contained dense retrieval system, these questions need to be answered.

In this paper, we evaluate the potential of dimension reduction in building lightweight dense retrieval systems. Specifically, as shown in Fig. 1, we build the DIRECTED, a simple but efficient dense retrieval system. DIRECTED utilizes a four-step pipeline, *encoding-compression-indexing-retrieval*, for retrieving document candidates. To implement our dimension reduction module, we employ conditional autoencoders (ConAE) [15] in our DIRECTED system, which is a state-of-the-art method.

Our experiments show the ConAE can broaden its effectiveness in different retrieval scenarios and also brings its dimension reduction effectiveness to different index building methods. Overall, DIRECTED is able to build efficient dense retrieval systems by reducing the retrieval latency by 96% and document index storage by 98%, while maintaining comparable retrieval performance with vanilla dense retrieval systems.

In conclusion, the contributions of this paper can be summarized as follows:

– DIRECTED is the first method to incorporate the dimension reduction stage in the retrieval system, which helps users to tailor their retrieval systems to effectively reduce the index storage or accelerate the retrieval process.
– Our evaluation results on different retrieval datasets show that the dimension reduction model has strong generalization capabilities and exhibits acceptable performance in 15 different types of retrieval tasks. There is no significant degradation of retrieval performance, when we conduct 6× dimension reduction of embeddings.

– Our experiments show that the generated embeddings of dense retrievers are redundant. The dimension reduction method can even slightly improve the retrieval performance when we reduce the embedding dimension from 768 to 256 and use the Product Quantization (PQ) as the indexer.

## 2 Related Work

Dense retrievers usually employs language models to encode queries and documents and map them in an embedding space for retrieval [9,11,13,32,33,36,38]. They usually use in-batch negatives [9], BM25 negatives [9], and hard negatives [32] to train language models to get the representations of queries and documents. To learn effective embeddings for queries and documents, dense retrievers usually maintain high-dimensional embeddings to fit training signals [18,26]. However, high-dimensional embeddings make the exact nearest neighbor search computationally expensive, leading to higher retrieval latency and larger index storage [23].

A widely adopted approach for building efficient dense retrieval systems is to utilize approximate nearest neighbor (ANN) search, which aims to find a compromise between search efficiency and retrieval accuracy [37]. Common ANN indexing techniques include clustering methods [7,21], Locality-Sensitive Hashing (LSH), and graph-based methods [20], all of which help create document indexes optimized for dense retrieval. The Binary Passage Retriever (BPR) [34] encodes documents with binary codes and trains both the binary codes and document embeddings through contrastive training. Other approaches, mainly centered around Product Quantization, attempt to directly learn the document index, either by optimizing it through contrastive learning [31,37] or by framing the index learning process as a teacher-student distillation task [30].

In high-dimensional spaces, the "curse of dimensionality" makes it impractical to search through all data points to find the nearest neighbors [5]. Recent research highlights that maintaining high-dimensional text embeddings is crucial for query and document encoders to effectively capture supervised signals [18]. However, efforts have begun to focus on reducing embedding dimensions to minimize redundancy and enhance efficiency [23]. The most straightforward approach to dimension reduction is to retain only a portion of the original high-dimensional embeddings [18,35]. For example, some methods use the first 128 dimensions to encode both queries and documents [35], while others apply principal component analysis (PCA) to preserve the principal dimensions and retain most of the information from the original embeddings [18]. In contrast to these unsupervised techniques, other work [15,18] introduces a supervised approach that leverages neural networks to compress high-dimensional embeddings into lower-dimensional spaces. These supervised models achieve better dimension reduction by preserving more information compared to unsupervised models. To further optimize query and document encoders, Ma et al. [18] continuously train dense retrievers with contrastive learning strategies, employing various negative sampling methods [9,32]. However, retraining all parameters of the query and document encoders requires significant computational resources to re-encode the documents. Conditional autoencoder [15] addresses this by fixing the query and document encoders and fine-tuning only a linear layer to project document embeddings into lower-dimensional space.

# 3    Methodology

In this section, we introduce the lightweight dense retrieval framework by using the **DI**mension **RE**du**CT**ion based ind**E**x buil**D**ing (DIRECTED) framework.

## 3.1    Lightweight Dense Retrieval Framework

In this part, we first introduce the dense retrieval framework and then describe the efficient retrieval pipeline (DIRECTED).

**Preliminaries of Dense Retrieval.** We briefly introduce how dense retrieval works during training and inference. Given a query $q$ and a collection of documents $\mathcal{D} = \{d_1, \ldots, d_j, \ldots, d_n\}$, the goal of the retrieval task is to find a subset of documents that are relevant to the query. Dense retrieval models [9,32,33] learn a tailored embedding space and model the relevance between queries and documents. They employ pretrained language models, to encode $q$ and $d$ into $K$-dimensional dense vectors, $h_q$ and $h_d$:

$$h_q = \text{BERT}(q), h_d = \text{BERT}(d). \tag{1}$$

Then we contrastively train query and document encoders to optimize embedding space by maximizing the retrieval probability $P(d^+|q, \{d^+\} \cup \mathcal{D}^-)$ of the relevant document $d^+$ [32,33]:

$$P(d^+|q, \{d^+\} \cup \mathcal{D}^-) = \frac{e^{f(h_q, h_{d+})}}{e^{f(h_q, h_{d+})} + \sum\limits_{d^- \in \mathcal{D}^-} e^{f(h_q, h_{d-})}}, \tag{2}$$

where $d^-$ is the negative document and $\mathcal{D}^-$ is the irrelevant document set. The irrelevant documents can be sampled from inbatch negatives, BM25 negatives, and hard negatives [9,32]. The relevance score $f(q, d)$ of $q$ and $d$ can be calculated with dot product:

$$f(h_q, h_d) = h_q \cdot h_d, \tag{3}$$

where the dimension of $h_q$ and $h_d$ are 768 for the BERT-base model.

**Efficient Retrieval with DIRECTED.** After learning an embedding space for dense retrieval, dense retrieval models [9,32,33] usually employ flat index to search the most relevant documents according to the query $q$ by sorting the similarity scores $f(h_q, h_d)$ of the query $q$ and document $d$. However, $h_d$ is usually high-dimensional, leading to longer retrieval latency and more index storage.

To build an efficient retrieval model for the high-dimensional document embeddings, we propose a new lightweight retrieval framework DIRECTED, which incorporates two modules including dimension reduction and efficient index building in the retrieval pipeline. Specifically, DIRECTED uses conditional autoencoder [15] to reduce the dimension of document embeddings (Sect. 3.2), and then employ attribute index building methods to conduct approximate nearest neighbor search (Sect. 3.3).

## 3.2 Dimension Compression with Conditional Autoencoder

We employ conditional autoencoder (ConAE) [15] to compress the high-dimensional embeddings $h_q$ and $h_d$ of both queries and documents to the low-dimensional embeddings $h_q^e$ and $h_d^e$. Conditional autoencoder is different from previous work [18] and fix the query and document encoders (Eq. 1).

**Encoder.** We first get the initial dense representations for query $q$ and document $d$ through dense retrievers. Then, these high-dimensional embeddings can be compressed into low-dimensional ones:

$$h_q^e = \text{Linear}_q(h_q); h_d^e = \text{Linear}_d(h_d), \qquad (4)$$

where $h_q^e$ and $h_d^e$ are $L$-dimensional embeddings of $q$ and $d$, which are respectively encoded by two different linear layers, $\text{Linear}_q$ and $\text{Linear}_d$. The dimension $L$ can be much lower than the dimension of initial representations.

Then, the KL divergence was used to regulate encoded embeddings to mimic the initial embedding distributions of queries and their top-ranked documents:

$$L_{KL} = \sum_q \sum_d P(d|q, \mathcal{D}_{\text{top}}) \cdot \log \frac{P(d|q, \mathcal{D}_{\text{top}})}{P_e(d|q, \mathcal{D}_{\text{top}})}, \qquad (5)$$

where $P_e(d|q, D_{\text{top}})$ is calculated with E.q. 2, using the encoded embeddings $h_q^e$ and $h_d^e$.

**Decoder.** The decoder module maps the encoded embeddings $h_q^e$ and $h_d^e$ into the original embedding space and aligns them with $h_q$ and $h_d$, aiming at optimizing encoder modules to maintain ranking features from $h_q$ and $h_d$ maximally.

Firstly, we use one linear layer and project $h_q^e$ and $h_d^e$ to $K$-dimensional embeddings, $\hat{h}_q$ and $\hat{h}_d$:

$$\hat{h}_q = \text{Linear}(h_q^e); \hat{h}_d = \text{Linear}(h_q^e). \qquad (6)$$

Then we respectively train the decoded embeddings $\hat{h}_q$ and $\hat{h}_d$ with the original frozen document and query embeddings of the retriever model to recover ranking features from vanilla retriever model and better align them with $h_q$ and $h_d$ in the original embedding space. The first loss $L_q$ is used to optimize the generated query representation $\hat{h}_q$:

$$L_q = \sum_q \sum_{d^+ \in \mathcal{D}^+} \frac{e^{f(\hat{h}_q, h_{d+})}}{e^{f(\hat{h}_q, h_{d+})} + \sum_{d^- \in \mathcal{D}^-} e^{f(\hat{h}_q, h_{d-})}}, \qquad (7)$$

and we can also optimize the generated document representation $\hat{h}_d$ with the second loss function $L_d$:

$$L_d = \sum_q \sum_{d^+ \in \mathcal{D}^+} \frac{e^{f(h_q, \hat{h}_{d+})}}{e^{f(h_q, \hat{h}_{d+})} + \sum_{d^- \in \mathcal{D}^-} e^{f(h_q, \hat{h}_{d-})}}, \qquad (8)$$

**Training Loss.** Finally, we train our conditional autoencoder model with the following loss $L$:

$$L = L_{KL} + \lambda L_q + \lambda L_d, \tag{9}$$

where $\lambda$ is a hyper-parameter to weight the autoencoder losses.

### 3.3 Efficient Vector Search with ANNs

In this subsection, we build index for the document $D = \{d_1, ..., d_n\}$ and introduce several widely used approximate nearest neighbor search methods as the plugin of the retrieval pipeline

**Inverted File Index (IVF)** [27] The training process of IVF contains two steps: 1) *Inverted Index Building.* IVF employs the $k$-means [19] method to cluster the document embeddings. It divides the document embeddings $h_d^e$ into $k$ embedding groups and uses the centroid representations of document embeddings $H_C^e = \{h_{c_1}^e, h_{c_2}^e, ..., h_{c_k}^e\}$ to represent the documents belong to different groups. The document centroid representations $H_C$ are used as the additional document index. 2) *Retrieval using IVF index.* During retrieval, we first calculate the similarity score $f(h_q^e, h_{c_i}^e)$ between query embedding $h_q^e$ and the $i$-th centroid embedding $h_{c_i}^e$. Then the $j$-th document group is selected, which has the highest similarity score with the user query:

$$c_j = \arg\max_i f(h_q^e, h_{c_i}^e). \tag{10}$$

Finally, the relevant documents are selected from the $j$-th document group. Ideally, these $n$ documents in the collection $D$ are uniformly divided into $k$ groups, thus, the similarity computation is called $n/k + k$ times, while $n$-time similarity computations on the flat index.

**Product Quantization (PQ)** [7] divides the dimensional reduced embedding $h_d^e \in R^L$ and $h_q^e \in R^L$ into $m$ subvectors:

$$h_d^e = h_d^e(1); h_d^e(2); ...; h_d^e(m); \tag{11}$$

$$h_q^e = h_q^e(1); h_q^e(2); ...; h_q^e(m), \tag{12}$$

where ; is the concatenation operation and $h_q^e(i) \in R^{L/m}$, $h_d^e(i) \in R^{L/m}$. Then PQ method employs $k$-means method to cluster the $i$-th subvectors of documents and represents these $i$-th subvectors with $k$ codebook embeddings $H_c^i = \{h_{c_1}^i, h_{c_2}^i, ..., h_{c_k}^i\}$. Finally, the similarity score $f(h_q^e, h_d^e)$ between query and document can be calculated:

$$f(h_q^e, h_d^e) = \sum_i h_q^e(i) \cdot h_d^e(i) \tag{13}$$

$$\approx \sum_i h_q^e(i) \cdot h_{c_*}^i. \tag{14}$$

$h_{c_*}^i$ is the codebook embedding among $H_c^i$, which has the smallest L2 distance with $h_d^e(i)$.

**PQ-STE** [12] utilize straight-through estimator [1] to select centroids and codebook vectors, which makes the computation process differentiable. Specifically, for each query $q$ and a collection of documents $D$, a trained dense retriever will give a teachers' score $s_{q,d}$ for each query-document pair:

$$s_{q,d} = f(h_q, h_d), \tag{15}$$

the softmax activation of the prediction can be computed as:

$$\sigma_{q,d} \leftarrow \frac{\exp\left(s_{q,d}\right)}{\sum_{d' \in D} \exp\left(s_{q,d'}\right)}, \tag{16}$$

For the quantization vectors, PQ-STE uses the same steps to compute the students' score $\tilde{s}_{q,d}$ and the softmax activation $\tilde{\sigma}_{q,d}$. Then, it minimizes the KL divergence between $\tilde{\sigma}_{q,d}$ and $\tilde{\sigma}_{q,d}$ to maintain the same ranking features as the original vector:

$$L = \sum_q \sum_d \sigma_{q,d} \cdot \log \tilde{\sigma}_{q,d}. \tag{17}$$

**IVFSQ** [40] combines IVF and SQ [17] to build the document index. This hybrid approach not only enhances retrieval efficiency but also reduces embedding dimensions, leading to significantly reduced retrieval latency and memory requirements, albeit with a minor trade-off in accuracy. As a result, it finds applicability in scenarios prioritizing deployment efficiency and rapid retrieval over absolute precision.

**Hierarchical Navigable Small World (HNSW)** [20] is a graph-based method that builds a hierarchical graph by treating embeddings as vertices structure, where each layer is an approximate *Small-World* network. Given a query, it searches over the graph by greedily selecting similar neighbors. Since the entire graph is connected, the query can move around between the different layers and thus find the eligible document embeddings extremely fast.

## 4  Experimental Methodology

In this section, we describes the datasets, baselines and implementation details of our experiments.

**Dataset.** MS MARCO [24] Passage Ranking dataset is one of the most prestigious in dense retrieval research, containing 8.8 million passages and 0.5 million queries from Bing's user search query logs and web documents retrieved. Specifically, we randomly sample 50,000 queries from the raw training set of MS MARCO as the dev set and use official dev sets as the test set. To further explore the impact of ConAE on out-of-domain performance with supervised dimension reduction, we evaluate the retrieval effectiveness of ConAE on the BEIR benchmark [28]. BEIR contains 9 heterogeneous search tasks covering 18 different datasets (15 of them are open source datasets), which is a recent standard benchmark for zero-shot dense retrieval.

**Metrics.** Normalized Discounted Cumulative Gain (nDCG) quantifies the effectiveness of a ranking algorithm by considering both the relevance of items and their positions in the list:

$$\text{nDCG@}k = \frac{1}{|\mathcal{Q}|} \sum_{q=1}^{|\mathcal{Q}|} \frac{\text{DCG}_q\text{@}k}{\text{IDCG}_q\text{@}k}, \tag{18}$$

where $k$ represents the length of the considered ranking list and $\mathcal{Q}$ represents the query set. $DCG@k$ is the calculation of discounted cumulative gain, and $IDCG@k$ is the ideal discounted cumulative gain under ideal ranking conditions. The calculation formula for IDCG@$k$ is:

$$\text{DCG}_q\text{@}k = \sum_{i=1}^{k} \frac{2^{rel_i} - 1}{\log_2(i+1)}, \tag{19}$$

where $rel_i$ is the graded relevance score for the $i$-th retrieved text, and IDCG@$k$ is the DCG value in the ideal case, where items are ranked based on relevance in descending order. MS MARCO also uses Mean Reciprocal Rank (MRR) and Recall as the primary evaluation metrics. The formula for calculating MRR is as follows:

$$\text{MRR} = \frac{1}{|\mathcal{Q}|} \sum_{q=1}^{|\mathcal{Q}|} \frac{1}{\text{rank}_q}, \tag{20}$$

where $rank_q$ indicates the ranking of the most relevant text in the top-k retrieval results. $Recall@k$ was designed to measure a retrieval system's ability to retrieve relevant documents within the top k successfully returned results. The formula to calculate Recall @k is as follows:

$$\text{Recall@}k = \frac{1}{|\mathcal{Q}|} \sum_{q=1}^{|\mathcal{Q}|} \frac{\#\,\text{retr}_{q,k}}{\#\,\text{rel}_q}, \tag{21}$$

where $\#\,\text{retr}_{q,k}$ denotes the number of relevant passages that query $q$ retrieved at top k positions by retriever, and $\#\,\text{retr}_q$ denotes the total number of relevant texts for query $q$.

**Implementation Details.** All of conditional autoencoder models are based on coCondenser [4]. For each query, we use one negative document to train it contrastively. We set the batch size to 128 during training. All models are implemented with PyTorch and tuned with Adam optimizer. The learning rates of all models are set to 1e-3. $\lambda$ of conditional autoencoder is set to 0.1. The number of centroids of IVF was 10,000, and the inner product was used as the quantizer method. We select data from 100 cluster centers for calculation while searching. For product quantization, different dimensions are divided into different subembeddings. The number of clustering centers is 256, and the subembedding dimension is kept as 4 when partitioning. For HNSW, the number of layers in the index is set to 200, the number of neighbors per vertex is 512, and the layer depth in the search is 128.

**Table 1.** Overall Performance on MSMARCO with Different Index Building Methods.

| Method | Size | SCR | Lant. | LCR | MRR@10 | NDCG@10 | Rec@1000 |
|---|---|---|---|---|---|---|---|
| Teacher | 26 GB | - | 341.14 | - | 0.3812 | 0.4460 | 0.9841 |
| + PQ | 1.6 G | 16.3× | 66.07 | 5.2× | 0.3426 | 0.4068 | 0.9780 |
| + PQ-STE | 1.6 G | 16.3× | 65.15 | 5.2× | 0.3753 | 0.4387 | 0.9823 |
| + IVF | 26 G | 1.0× | 8.90 | 38.3× | 0.3602 | 0.4198 | 0.9170 |
| + IVFSQ | 13 G | 2.0× | 2.51 | 136.0× | 0.3599 | 0.4198 | 0.9174 |
| + HNSW | 60 G | 0.4× | 1.45 | 235.4× | 0.3791 | 0.4436 | 0.9784 |
| ConAE-256 | 8.5 G | 3.1× | 114.45 | 3.0× | 0.3802 | 0.4451 | 0.9836 |
| + PQ | 541 M | 49.2× | 19.08 | 17.9× | 0.3521 | 0.4168 | 0.9789 |
| + PQ-STE | 541 M | 49.2× | 18.99 | 18.0× | 0.3572 | 0.4198 | 0.9783 |
| + IVF | 8.6 G | 3.0× | 2.50 | 135.5× | 0.3581 | 0.4177 | 0.9189 |
| + IVFSQ | 4.3 G | 6.0× | 0.90 | 377.8× | 0.3567 | 0.4153 | 0.9192 |
| + HNSW | 43 G | 0.6× | 0.89 | 384.6× | 0.3709 | 0.4341 | 0.9696 |
| ConAE-128 | 4.3 G | 6.0× | 62.44 | 5.5× | 0.3632 | 0.4257 | 0.9788 |
| + PQ | 271 M | 98.2× | 9.31 | 36.6× | 0.3251 | 0.3858 | 0.9718 |
| + PQ-STE | 271 M | 98.2× | 9.68 | 35.2× | 0.3338 | 0.3945 | 0.9694 |
| + IVF | 4.3 G | 6.0× | 1.34 | 254.6× | 0.3406 | 0.3984 | 0.9118 |
| + IVFSQ | 2.2 G | 11.8× | 0.56 | 609.2× | 0.3412 | 0.3988 | 0.9108 |
| + HNSW | 39 G | 0.5× | 0.78 | 438.5× | 0.3587 | 0.4207 | 0.9682 |
| ConAE-64 | 2.2 G | 11.8× | 36.03 | 9.5× | 0.3165 | 0.3717 | 0.9464 |
| + PQ | 138 M | 195.8× | 8.20 | 41.6× | 0.2669 | 0.3178 | 0.9238 |
| + PQ-STE | 138 M | 195.8× | 6.99 | 48.8× | 0.2890 | 0.3431 | 0.9212 |
| + IVF | 2.2 G | 11.8× | 0.49 | 696.2× | 0.2994 | 0.3498 | 0.8643 |
| + IVFSQ | 1.2 G | 21.7× | 0.40 | 852.9× | 0.2996 | 0.3500 | 0.8646 |
| + HNSW | 37 G | 0.7× | 0.66 | 513.8× | 0.3150 | 0.3704 | 0.9437 |

## 5    Experimental Results

In this section, we first explore the effectiveness of our DIRECTED method. And then we further evaluate the generalization and compatibility of conditional autoencoders (ConAE) with different retrieval scenarios and index building methods.

### 5.1    Overall Performance

The retrieval performance of different models is shown in Table 1, which include size compression ratio (SCR) and latency compression ratio (LCR) to reflect improvements in efficiency. Superior results are achieved with our pipeline. DIRECTED shows strong effectiveness in building an efficient retrieval system.

Specifically, DIRECTED can compress the index size by maximum of 195.8× with the 64-dim reduction plug-in (ConAE-64) and the efficient indexing plug-in of the PQs.

ConAE-64 + IVFSQ achieves a maximum acceleration ratio of 852.9 × for retrieval latency. The ultimate performance degradation is 11.43%, which is still acceptable.

A further novel finding is that DIRECTED can maintain the compatible retrieval performance by 128-dim dimension reduction plug-ins (ConAE-128) and IVF method. It significantly diminishes the index size to 17% of the original size and achieves a retrieval latency reduction of 254.6 times compared to 768-dim dense vectors. Remarkably, MRR@10 on MSMARCO reduced by only 4%. Such results demonstrate our DIRECTED's effectiveness as an efficient dense retrieval system. It is worth noting that the size of index obtained by ConAE-256 with IVFSQ is similar to using ConAE-128 only. Although the ConAE-128 is more efficient than the ConAE-256 with IVFSQ, the retrieval speed is much lower. The same pattern appears for the combination of ConAE-128 + IVFSQ with the flat index of ConAE-64, which suggests that the combination of dimension reduction and efficient indexing may be complementary. The collaboration between the two plug-ins can break through the efficiency limit of a single method and realize a more efficient retrieval system.

We then compare the relationship between retrieval dimension, latency, and accuracy on the MS-MARCO dataset, as shown in Fig. 2. It can be observed that most ANN methods exhibit antagonism between retrieval latency and retrieval accuracy, i.e., the decreasing trend of retrieval latency is consistent with retrieval accuracy. Thus, it's hard to achieve both retrieval effectiveness and retrieval efficiency simultaneously in one retrieval system. Existing jointly embedding and index learning methods usually face the unstable training problem due to the conflict training objects [12,30,38]. However, we may be able to find a more balanced point between retrieval latency and accuracy based on these analyses. When DIRECTED used ConAE to reduce the dimension to $d = 256$, it always achieved a more delicate balance between retrieval latency and retrieval accuracy. This is mainly reflected in the slope of the broken line: the retrieval latency decreases sharply, while the retrieval accuracy decreases more slowly. And when $d = 128$, the decrease of latency and precision is almost synchronized. This means that at $d = \{256, 128\}$, DIRECTED can guarantee a reasonable retrieval accuracy with sufficient efficiency. However, when the dimension is reduced to $d = 64$, the decrease in retrieval accuracy is dramatic with or without the efficient indexing plug-in. This is perhaps due to the loss of ranking features caused by the 64-dim vector during the dimension reduction process, which leads to suboptimal retrieval accuracy.

An anomaly worth exploring is that the combination of ConAE-256 + PQ breaks the adversarial relationship between retrieval latency and accuracy. It improves the retrieval accuracy while reducing the latency. Based on this finding, we hypothesize that the 768-dim vector contains more redundant information, and the codewords mostly come from redundant sub-embeddings during unsupervised clustering, which results in the poor effect of Teacher-768 + PQ. Our dimension reduction plug-in reduces the redundancy of the representations while reducing the dimension and achieving higher retrieval performance. In contrast, PQ-STE supervised the optimization of the codebook parameters by using the similarity scores of query-document pairs as the labeling formulation optimization target, resulting in the clustering of codewords into information-rich sub-embeddings, so the performance of Teacher-768 + PQ-STE is in line with the general pattern.

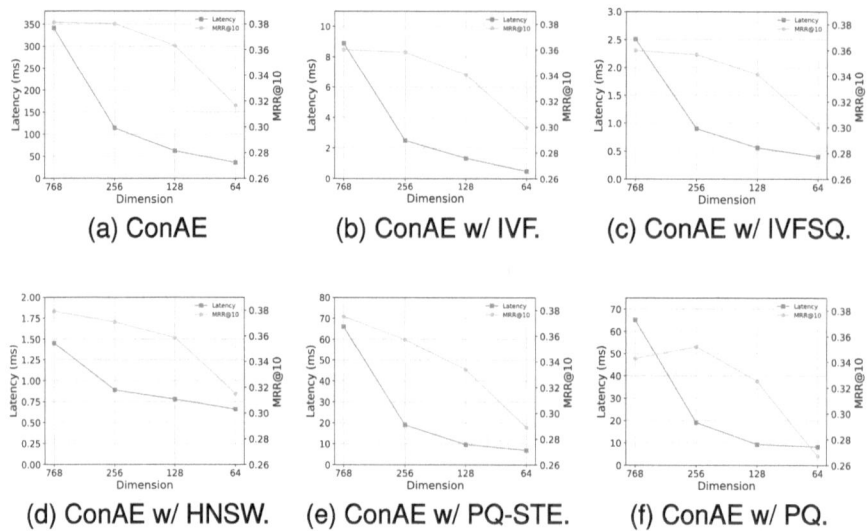

**Fig. 2.** Performance-latency Curves Using Different ANN Methods. The orange and blue lines represent the performance loss and latency degradation with decreasing dimensionality, respectively. (Color figure online)

### 5.2    Generalization and Compatibility of the Dimension Reduction Model

We conduct different levels of dimension reduction, from $d = 256$ to $d = 64$, and then evaluate the domain adaption ability of ConAE by testing the retrieval effectiveness on the BEIR dataset.

Table 2 shows the zero-shot performance of our dimension reduction models across different dimension levels on BEIR benchmark; the numbers in the $\Delta$ column represent the differences in the performances of our models at various dimensions levels compared to the 768-dim original embeddings. Overall, our zero-shot experiments demonstrate that ConAE can maintain its effectiveness in the retrieval tasks of other domains by achieving comparable retrieval performance and significantly reducing the retrieval latency and index storage by dimension reduction. In addition, after reducing the dimension, the retrieval performance on out-domain data decreases more than on in-domain data, showing that the generalization ability will degenerate due to the fewer parameters.

We then notice that when $d = 256$, the dimension reduction plug-in yields results similar to those of the original 768-dim coCondenser (average performance reduction of 1%). For some tasks, the performance of the model is even slightly improved when the dimension was reduced from 768 to 256. Further, when we downscale the vector to $d = 128$, the performance of the model decreases from 4.96% to 19.19% for each dataset of BEIR compared to the original embedding. This result looks acceptable, considering that the dimension of the vector has been compressed by a factor of 6x compared to the original embedding. However, by comparing the results of the 64-dimensional model, we speculate that $d = 128$ may be a balance between model capacity and effectiveness.

**Table 2.** Retrieval Performance with Dimension Reduction Model (ConAE [15]) on Different Domains. The numbers in $\Delta$ column denote the performances of different reduced dimensions vs 768 dimensional embeddings.

| Dim.($\rightarrow$) Dataset($\downarrow$) | 768 | 256 | $\Delta$ | 128 | $\Delta$ | 64 | $\Delta$ |
|---|---|---|---|---|---|---|---|
| MS MARCO | 0.3812 | 0.3758 | +0.54% | 0.3632 | −1.80% | 0.3165 | −6.47% |
| TREC-COVID | 0.6758 | 0.6484 | −2.74% | 0.6140 | −6.18% | 0.4407 | −23.51% |
| NFCorpus | 0.3032 | 0.3040 | **+0.08%** | 0.2864 | −1.68% | 0.2271 | −7.61% |
| NQ | 0.3908 | 0.3875 | −0.33% | 0.3615 | −2.93% | 0.2871 | −10.37% |
| HotpotQA | 0.5312 | 0.5064 | −2.48% | 0.4460 | −8.52% | 0.2407 | −29.05% |
| FiQA-2018 | 0.2851 | 0.2802 | −0.49% | 0.2599 | −2.52% | 0.1898 | −9.53% |
| ArguAna | 0.2704 | 0.2672 | −0.32% | 0.2506 | −1.98% | 0.1962 | −7.42% |
| Touché-2020 | 0.2652 | 0.2640 | −0.12% | 0.2423 | −2.29% | 0.2119 | −5.33% |
| CQADupStack | 0.3328 | 0.3235 | −0.93% | 0.2942 | −3.86% | 0.2115 | −12.13% |
| Quora | 0.8636 | 0.8578 | −0.58% | 0.8448 | −1.88% | 0.7788 | −8.48% |
| DBPedia | 0.3481 | 0.3364 | −1.17% | 0.3030 | −4.51% | 0.1886 | −15.95% |
| SCIDOCS | 0.1216 | 0.1140 | −0.76% | 0.1048 | −1.68% | 0.0629 | −5.87% |
| FEVER | 0.6304 | 0.6103 | −2.01% | 0.5565 | −7.39% | 0.3354 | −29.50% |
| Climate-FEVER | 0.1750 | 0.1777 | **+0.27%** | 0.1490 | −2.60% | 0.0798 | −9.52% |
| SciFact | 0.5322 | 0.5087 | −2.35% | 0.4625 | −6.97% | 0.3543 | −17.79% |
| **AVG.** | 0.4071 | 0.3975 | −0.96% | 0.3692 | −3.79% | 0.2748 | −13.24% |

When $d = 64$, the model has a very significant degradation relative to the original embedding on all datasets, which may due to the lack of capacity of embedding to save effective features.

## 6 Conclusion

This paper further explore the potential of building a lightweight dense retrieval system by combining the dimension reduction in the retrieval pipeline. Our results demonstrates the value of the dimension reduction stage in building retrieval systems. With only a 3% decrease in retrieval performance, it achieves a 3% reduction in index storage and a 5% decrease in retrieval latency, making it feasible to deploy dense retrieval systems on edge devices.

**Acknowledgments.** This work is partly supported by the Natural Science Foundation of China under Grant (No. 62206042), the Joint Funds of Natural Science Foundation of Liaoning Province (No. 2023-MSBA-081), and the Fundamental Research Funds for the Central Universities under Grant (No. N2416012).

# References

1. Bengio, Y., Léonard, N., Courville, A.: Estimating or propagating gradients through stochastic neurons for conditional computation. arXiv preprintarXiv:1308.3432 (2013)
2. Chen, J., et al.: BGE M3-Embedding: multi-lingual, multi-functionality, multi-granularity text embeddings through self-knowledge distillation. arXiv preprint arXiv:2402.03216 (2024)
3. Gao, L., Callan, J.: Condenser: a pre-training architecture for dense retrieval. In: Proceedings of EMNLP, pp. 981–993 (2021)
4. Gao, L., Callan, J.: Unsupervised corpus aware language model pre-training for dense passage retrieval. In: Proceedings of ACL, pp. 2843–2853 (2022)
5. Indyk,P., Motwani, R.: Approximate nearest neighbors: towards removing the curse of dimensionality. In: Proceedings ACM STOC, pp. 604–613 (1998)
6. Izacard, G., Grave, É.: Leveraging passage retrieval with generative models for open domain question answering. In: Proceedings of EACL, pp. 874–880 (2021)
7. Jégou, H., Douze, M., Schmid, C.: Product quantization for nearest neighbor search. IEEE Trans. Pattern Anal. Mach. Intell. **33**(1), 117–128 (2011)
8. Johnson, J., Douze, M., Jégou, H.: Billion-scale similarity search with GPUs. IEEE Trans. Big Data **7**(3), 535–547 (2019)
9. Karpukhin, V., et al.: Dense passage retrieval for open-domain question answering. In: Proceedings of EMNLP, pp. 6769–6781 (2020)
10. Kwiatkowski, T., et al.: Natural questions: a benchmark for question answering research. Trans. Assoc. Comput. Linguist. **7**, 452–466 (2019)
11. Lewis, M., et al.: Pre-training via paraphrasing. In: Proceedings of NeurIPS (2020)
12. Li, C., et al.: LibVQ: a toolkit for optimizing vector quantization and efficient neural retrieval. In: Proceedings of SIGIR, pp. 3095–3099 (2023)
13. Li, Y., et al.: More robust dense retrieval with contrastive dual learning. In: Proceedings of SIGIR ICTIR, pp. 287–296 (2021)
14. Liu, Z., et al.: Adapting open domain fact extraction and verification to COVID-fact through in-domain language modeling. In: Findings of ACL, pp. 2395–2400 (2020)
15. Liu, Z., et al.: Dimension reduction for efficient dense retrieval via conditional autoencoder. In: Proceedings of EMNLP, pp. 5692–5698 (2022)
16. Liu, Z., et al.: Fine-grained fact verification with kernel graph attention network. In: Proceedings of ACL, pp. 7342–7351 (2020)
17. Lloyd, S.: Least squares quantization in PCM. IEEE Trans. Inf. Theory **28**(2), 129–137 (1982)
18. Ma, X., et al.: Simple and effective unsupervised redundancy elimination to compress dense vectors for passage retrieval. In: Proceedings of the EMNLP, pp. 2854–2859 (2021)
19. MacQueen, J., et al.: Some methods for classification and analysis of multivariate observations. In: Proceedings of the Fifth Berkeley Symposium on Mathematical Statistics and Probability, vol. 1, pp. 281–297. Oakland, CA, USA (1967)
20. Malkov, Y.A.: Efficient and robust approximate nearest neighbor search using hierarchical navigable small world graphs. IEEE Trans. Pattern Anal. Mach. Intell. **42**(4), 824–836 (2018)
21. McKusick, M.K., et al.: A fast file system for UNIX. ACM Trans. Comput. Syst. **2**(3), 181–197 (1984)
22. Meiser, S.: Point location in arrangements of hyperplanes. Inf. Comput. **106**(2), 286–303 (1993)
23. Min, S., et al.: NeurIPS 2020 efficientQA competition: Systems, analyses and lessons learned. In: Proceedings of the NeurIPS 2020 Competition and Demonstration Track, pp. 86–111 (2021)

24. Nguyen, T., et al.: MS MARCO: a human generated machine reading comprehension dataset. In: Proceedings of NIPS, vol. 1773 (2016)
25. Ni, J., et al.: Large dual encoders are generalizable retrievers. In: Proceedings of EMNLP, pp. 9844–9855 (2022)
26. Reimers, N., Gurevych, I.: The curse of dense low-dimensional information retrieval for large index sizes. In: Proceedings of ACL, pp. 605–611 (2021)
27. Salton, G., Fox, E.A., Wu, H.: Extended Boolean information retrieval. Commun. ACM **26**(11), 1022–1036 (1983)
28. Thakur, N., et al.: BEIR: a heterogeneous benchmark for zero-shot evaluation of information retrieval models. In: NeurIPS Datasets and Benchmarks Track (2021)
29. Xiao, S., et al.: C-Pack: packaged resources to advance general Chinese embedding. arXiv preprint arXiv:2309.07597 (2023)
30. Xiao, S., et al.: Distill-VQ: learning retrieval oriented vector quantization by distilling knowledge from dense embeddings. In: Proceedings of SIGIR, pp. 1513–1523 (2022)
31. Xiao, S., et al.: Matching-oriented embedding quantization for ad-hoc retrieval. In: Proceedings of EMNLP, pp. 8119–8129 (2021)
32. Xiong, L., et al.: Approximate nearest neighbor negative contrastive learning for dense text retrieval. In: Processing of ICLR (2020)
33. Xiong, W., et al.: Answering complex open-domain questions with multi-hop dense retrieval. In: Processing of ICLR (2020)
34. Yamada, I., Asai, A., Hajishirzi, H.: Efficient passage retrieval with hashing for open-domain question answering. In: Proceedings of ACL, pp. 979–986 (2021)
35. Yang, S., Seo, M.: Designing a minimal retrieve-and-read system for open-domain question answering. In: Proceedings of NAACL-HLT, pp. 5856–5865 (2021)
36. Yu, S., et al.: Few-shot conversational dense retrieval. In: Proceedings of SIGIR, pp. 829–838 (2021)
37. Zhan, J., et al.: Jointly optimizing query encoder and product quantization to improve retrieval performance. In: Proceedings of CIKM, pp. 2487–2496 (2021)
38. Zhan, J., et al.: Optimizing dense retrieval model training with hard negatives. In: Proceedings of SIGIR, pp. 1503–1512 (2021)
39. Zhao, W.X. et al.: Dense text retrieval based on pretrained language models: a survey. arXiv preprint arXiv:2211.14876 (2022)
40. Zobel, J., Moffat, A.: Inverted files for text search engines. ACM Comput. Surv. **38**(2), 6 (2006)

# Relation Extraction Model Based on Overlap Rules and Abductive Learning

Zihui Wei[1], Yijia Zhang[2(✉)], Mingyu Lu[1], and Hongfei Lin[3]

[1] School of Artificial Intelligence, Dalian Maritime University, Dalian 116024, China
{1120220646,lumingyu}@dlmu.edu.cn
[2] School of Information Science and Technology, Dalian Maritime University,
Dalian 116024, China
zhangyijia@dlmu.edu.cn
[3] School of Computer Science and Technology, Dalian University of Technology,
Dalian 116024, China
hflin@dlut.edu.cn

**Abstract.** The exponential growth of unstructured text data on the internet presents a significant challenge in extracting entity relation entity triplets. Complex overlaps and trans-missions of relations between entities within the same text paragraph further exacerbate this challenge, often leading to contradictory results in extraction models. In this paper, we propose an innovative approach, the Relation Extraction Model based on Overlap Rules and Abductive Learning (ORABL), to address these issues. ORABL integrates an initial classifier with a rule module that identifies and corrects contradictions in the initial predictions. This iterative process continues until the model no longer produces contradictory conclusions, thereby improving the consistency of the extracted relations with natural language habits. Through extensive experimentation, we demonstrate that ORABL outperforms state-of-the-art models in various scenarios, including single entity overlap (SEO), entity pair overlap (EPO), and Subject Object Overlap (SOO). Our results underscore the effectiveness of ORABL in handling the complexities of relation extraction from unstructured text data. The results on FewRel indicate that it also has certain potential in solving the few-shot problem.

**Keywords:** Relational Triple Extraction · PLM · Abductive Learning

## 1 Introduction

Relational triple extraction (RTE) stands as a crucial algorithm in the automated construction of extensive knowledge bases, involving the joint identification of entities and their semantic relations from unprocessed texts in a seamless manner [1]. Early methodologies approached RTE tasks sequentially, dividing the process into two distinct phases where entities were initially identified, followed by the assignment of relations to each extracted entity pair [2]. Nonetheless, these

X. He et al. (Eds.): CCIR 2024, LNCS 15418, pp. 80–93, 2025.
https://doi.org/10.1007/978-981-96-1710-4_7

methods often fail to capture the implicit correlations between these segregated tasks, resulting in propagated errors [3].

In contrast, recent scholarly endeavors [4] have directed their focus towards jointly extracting $< subject - relation - object >$ triples in an end-to-end fashion. For instance, Wei [5] proposed a cascaded network that initially identified subjects and subsequently recognized corresponding objects for relations. Zheng [6] deconstructed RTE into three subtasks. Meanwhile, Wang [7] and Shang [8] conducted relational triple extraction in a single stage to alleviate exposure bias. Despite these advancements, prevalent approaches often disregard the rich informative correlations existing between entities and relations.

Unirel [9], on the other hand, advocates for a comprehensive unified relation modeling approach to achieve more effective context representation and information exchange. Despite these strides, extant works still overlooked the intrinsic logic embedded within relational triplets as structured data, consequently rendering the model vulnerable in handling intricate environments.

To elaborate, contemporary methods exhibit certain shortcomings: they concentrate on the interaction of information within the input text but neglect the inherent logic underlying each triplet. The reliance on deep learning black boxes can cause intricate overlapping issues to be prone to confusion.

In this paper, we propose a stepwise extraction model grounded in logic module perception, denoted as ORABL. We embrace the Abductive Learning framework and employ Relation Overlap Rules as the Abductive module to augment classifier training. We fully exploit the advantages offered by triples as structured information and enhance model efficacy through the incorporation of mutual exclusion and inclusion relations among triples. Owing to the utilization of the Abductive learning [10] framework, the model also attains a certain level of interpretability and adeptness in handling few-shot scenarios.

## 2  Related Works

In contemporary research, Relational Triple Extraction (RTE) has emerged as a focal point, with a pronounced trend towards the joint extraction of entities and relations. Initially, feature-based models dominated the landscape, necessitating intricate feature engineering and relying extensively on natural language processing (NLP) tools [3,11,12]. However, the advent of neural network-based joint models marked a significant shift, eliminating the need for manually crafted features. Miwa and Bansal [4] proposed parameter sharing as a means to facilitate the simultaneous learning of entities and relations, while Zheng [13] reframed RTE as a sequence tagging problem, amalgamating entities and relation annotations.

Despite these advancements, prevailing models often grapple with intricate scenarios characterized by overlapping regional triples, wherein entities or entity pairs are shared (SingleEntityOverlap, SEO, or EntityPairOverlap, EPO). Generative models have emerged as a potential solution, treating triples as token sequences to address these challenges [14,15]. However, methods aiming to extract triples in a single stage encounter hurdles due to the expansive prediction space [7,8,16].

Alternative approaches decompose RTEs into subtasks, yet they often struggle to effectively capture interactions between these subtasks, resulting in cascading errors [5,6,17–19]. Despite attempts to unify entity-entity and entity-relation interactions, such models typically employ separate modules for modeling, thereby limiting their efficacy [20]. EmRel, proposed by Xu [21], introduced relation representation explicitly to exploit interactions across relations, entities, and contexts but encountered challenges owing to heterogeneity between entities and relation embeddings.

Recent methodologies, such as prompt-tuning, have transformed relation extraction into masked language modeling, albeit with a primary focus on sentence-level relation classification, often neglecting entity-relation correlations [18,22,23]. Additionally, SSAN [24] delved into the self-attention layer within transformer architectures, primarily addressing document-level RE tasks by modeling structural interactions.

Abductive learning [10] introduces a novel machine learning framework that integrates machine learning and logical reasoning. However, owing to its reliance on structured knowledge bases and limited exploration of downstream tasks, only a small fraction of work has delved into the field of legal trials [25] in recent years.

In contrast to existing approaches, our study endeavors to address the challenge of underutilizing information by exploring mutual exclusion and inclusion phenomena among triples, thereby enhancing the model's ability to handle more complex tasks.

## 3   Methodology

In this section, we introduce the ORABL framework. We first present the problem setting and then provide an overview of the framework, followed by the optimization details. Figure 1 shows that our framework consists of a main classifier f and an Abductive module.

### Problem Formulation

Given a sentence $S = \{x_1, x_2, ......x_N\}$ with N tokens, the goal of joint relational triple extraction is to identify all possible triples $T = [(sub_l, rel_l, obj_l)]_{l=1}^{L}$ from S, where $sub_l, rel_l, obj_l$ represent the subject, the object, and their relation, respectively, and L is the number of triples. The subject and object are entity mentions $E = \{e_1, e_2......e_k\}$ from sentence S, where k is the number of entities. The relation is from the predefined relation set $R = \{rel_1, rel_2......rel_M\}$ with M types. Note that entities and relations might be shared among triples.

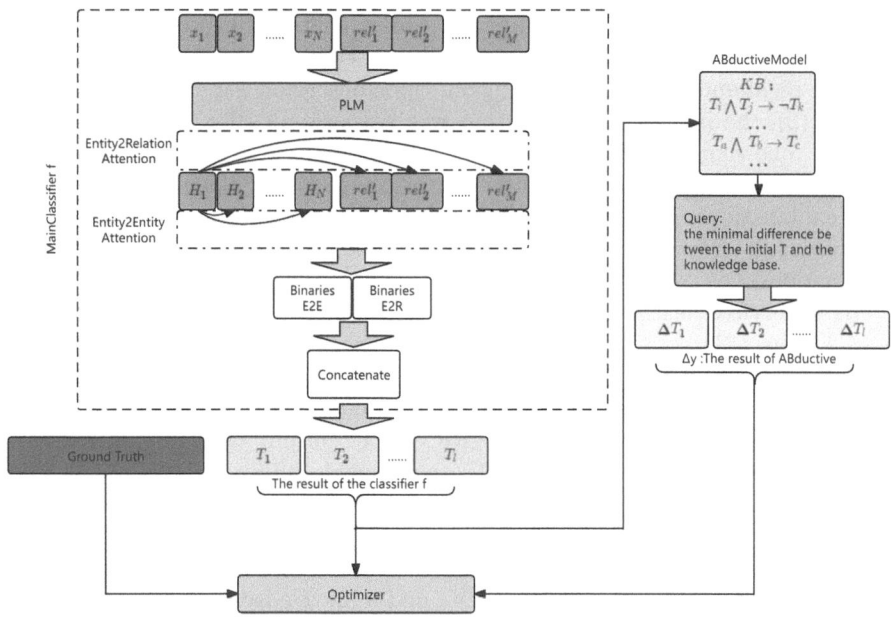

**Fig. 1.** Main framework of ORABL.

## 3.1 Abductive Learning

Abductive learning [10](ABL)is an innovative framework that harmonizes two foundational AI paradigms-machine learning and logical reasoning-to yield a symbiotic relation. Within the ABL, the machine learning model is trained to discern fundamental logical facts from the raw data. Concurrently, logical abduction leverages symbolic domain knowledge to rectify inaccuracies in perceived facts, thus enhancing the capabilities of machine learning models.

Abduction, also known as abductive reasoning, stands as one of the three primary forms of logical inference, alongside deduction and induction. While deduction entails deriving specific conclusions from general principles, and induction involves deriving general principles from specific observations, abduction diverges by crafting a foundational hypothesis to elucidate observed phenomena.

For clarity, this paper adopts the following notation for logical symbols: "$\neg$" represents negation (not); "$\wedge$" denotes conjunction (and); "$\vee$" signifies disjunction (or); and "$\leftarrow$" indicates implication, wherein the satisfaction of the condition on the right implies the validity of the consequent on the left. For instance, consider the following logical rules:

$$wetgrass \leftarrow rainlastnight \vee sprinklerwason, \qquad (1)$$

$$wetshoes \leftarrow wetgrass, \qquad (2)$$

$$false \leftarrow rainlastnight \wedge sprinklerwason, \qquad (3)$$

The first two formulas elucidate the causes contributing to the wetness of grass and shoes, while the final formula delineates the mutual exclusivity between rain last night and the sprinkler being on. Upon observing wet shoes, as indicated by rule (2), the inference dictates that wet grass should also hold true. Subsequently, following the logical progression, in accordance with rule (1), both the occurrence of rain last night and the activation of the sprinkler emerge as plausible explanations. However, should the absence of rain last night be further observed, rule (3) delineates that the sprinkler being on stands is the sole viable explanation.

In supervised learning, we are given labeled sentences: $SS = \{S_1, S_2 \ldots S_i \ldots\}$, where $Si = \{x_{i1}, x_{i2} \ldots x_{iN}\}$, and x is the word token. The corresponding ground truth label $Y = \{y_1, y_2, \ldots, y_i \ldots\}$, where $yi = \{T_{i1}, T_{i2} \ldots T_{iL}\}$, and T is the relation triplet present in the sentence. The task is to learn a function $f : X \rightarrow Y$, which would give correct output over unseen data. In SSL(Self-Supervised Learning) [26,27], we provide additional unlabeled data $X_u = \{x_{l+1}, x_{l+2}, \ldots, x_{l+u}\}$, usually $l \ll u$. The unlabeled data are utilized to improve performance of function f. In regard to the setting of ORABL, the input data still contain labeled data $X_l$ and their ground-truth label $Y_l$.

However, we have unlabeled data $X_u$ and a knowledge base KB. An incomplete KB consists of a number of first-order symbolic rules. The goal of learning is to utilize all labeled and unlabeled data to train a classifier with the help of a KB.

$$L_{all} = L_l + \alpha(t)L_u - \beta Con_{KB} \tag{1}$$

$$L_l = \frac{1}{N}\Sigma_{i=1}^{N_l} L(y_i, f_i) \tag{2}$$

$$L_u = \Sigma_{i=1}^{N_u} L(\Delta(y_i), f_i') \tag{3}$$

$$Con_{KB} = \Sigma_{i=1}^{N_u} Con(\Delta(y_i), KB) \tag{4}$$

where $N_l$ is the number of labeled data, $N_u$ for unlabeled data, $F_i$ is the function's output of labeled data, $y_i$ is the ground truth label $f_{i0}$ for unlabeled data, $y_{0i}$ is the pseudolabel, $\beta(y_{0i})$ is the revised pseudo-label by logical abduction process $\Delta$, which will be explained later. L is the loss function, and $\alpha(t)$ and $\beta$ are balancing coefficients. Con represents a function that outputs how consistent $\Delta(y_{0i})$ is with the KB.

Equation (2) is the loss of the labeled data. Equation (3) is the loss on the revised pseudolabels of unlabeled data. The first two terms are similar to the semisupervised loss function in pseudolabel. Equation (4) indicates the consistency between the revised pseudolabels and the knowledge base (the more consistency there is, the greater its value).

Our classifier conceptualizes the task of relation extraction as a series of multiobjective binary classification challenges, wherein each conceivable combination is independently evaluated for validity. Typically, only a limited number of labels are activated by a given input, leading the model to predominantly generate "negative" outcomes. To address this issue, we employ the Focus Loss function [28].

$$L = \begin{matrix} -(1-f)log(f), & Y = 1 \\ -flog(1-f), & Y = 0. \end{matrix} \qquad (5)$$

---

**Algorithm 1: ABL**

---

**Input**: input Labeled data and their labels $X_l, Y_l$;
Unlabeled data $X_u$;
Knowledge base KB
**Output**: output Function f
1   $f \leftarrow TrainModel(X_l, Y_l)$
2   while t < turn limit do
3   $Y_u \leftarrow f(X_u)\#Generate pseudo - labels : Y_u$
4   $\Delta(Y_u) \leftarrow Abduce(KB, Y_u)\#Revise pseudo - labels$
5   $f \leftarrow \varphi(f, X_u, \Delta(Yu), X_l, Y_l)\#Update function f$
6   $t \leftarrow t + 1$
7   end while

---

Algorithm 1 and Fig. 2 depict the main processes of ABproductive learning

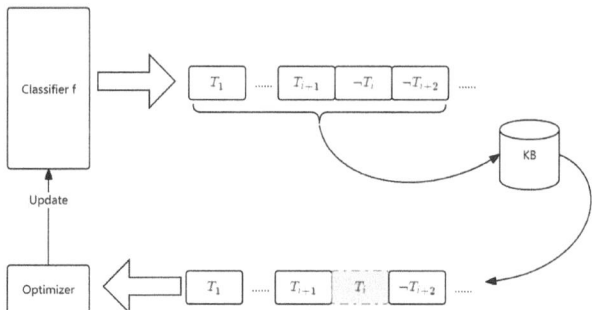

**Fig. 2.** Framework of ABL.

## 3.2   Classifier F

In our methodology, we commence by translating the relations delineated in the schema into natural language expressions. This transformation enables us to encapsulate richer relational information by converting the relation labels into descriptive statements. Subsequently, the converted re lational statements undergo a transfer process wherein they are mapped to an input embedding sequence through a search operation conducted within an embedding table. Following this, the synthesized connection between the input sentence and the natural language rendition of the relation is fed into a transformer-based pretrained language model (PLM) encoder. Specifically, we leverage BERT [29] as our chosen PLM. The input data are then converted into an input embedding sequence via another lookup operation within the embedding table.

$$H = Concat(H_s, H_r) \tag{6}$$

$$H_s = Emb(S) = Emb[x_1, x_2 \dots x_N] \tag{7}$$

$$H_r = Emb(R') \tag{8}$$

$$R' = EECT(R) \tag{9}$$

where Concat() represents the concatenation operation and Emb() is the embedding table of the pretrained model. EECT is an Eng-Eng comparison table used to transform relations into descriptive statements with complete information.

After acquiring the input embeddings, the encoder leverages the self-attention mechanism to capture the interactions among individual input words. Specifically, Transformer-based Pretrained Language Models (PLMs) are composed of stacked Transformer layers [30], layer incorporates multiple attention heads. Each attention head conducts three distinct linear transformations on the input embeddings H, yielding query (Q), key (K), and value (V) vectors, respectively. Subsequently, the attention weights between all pairs of words are computed through the softmax-normalized dot product of Q and K, which are then combined with V as follows:

$$Attention(Q, K, V) = softmax(\frac{QK^T}{\sqrt{d_h}})V \tag{10}$$

After the attention calculation, we obtained a set of highly correlated entity entities, as well as entity-relation binaries. These pairs will be preliminarily concatenated according to algorithm 2.

---

**Algorithm 2:** Concatenate

---

**Input:** $B_{e2e} = [(e_i, e_j)]_{all}$,
$B_{e2r} = [(e_i, r_m)]_{all}$,
$B_{r2e} = [(r_m, e_i)]_{all}$
**Output:** $T = [(sub_l, rel_l, obj_l)]$
1  initialize $T = \emptyset$
2  while all do
3  if $(e_i, e_j) \wedge (e_i, r_m) \wedge (r_m, e_j)$ then
4  $t = (e_i, r_m, e_j)$
5  end if
6  T=Concat(T,t)
7  end while

---

Then, these triples are input into the Abductive module for further correction, generating the correct triples.

## 4  Experiments

### 4.1  Datasets and Evaluation

We evaluated the proposed method on two widely used benchmark datasets, namely, the NYT [31] and WebNLG [32] datasets. The performance of the model was tested on FewRel under few-shot conditions. The NYT dataset poses a challenge similar to that of the TACRED dataset [33]. The SemEval dataset comprises a total of 8,000 training instances and 2,717 testing instances. For experimentation, the training set is divided into two subsets, with 7500 instances allocated for training and 500 instances allocated for development. Standard training, development, and testing splits provided by Zhang et al. [33] are employed for TACRED. To investigate the specific performance of the ABL framework, we designed a few shot experiments using the FewRel [34] dataset, which contains 100 categories and 70,000 instances. Our evaluation is based on standard data splitting, reporting the standard micro F1-score and MCC. Through observation, we find that the MCC [35]effectively gauges model superiority over simple probability predictions in multiclass tasks, thereby revealing the influence of the long-tail distribution within the training data on model learning. The higher the MCC is, the more balanced the performance of the model.

Our model is implemented using PyTorch. Parameter optimization is conducted via Adam optimization [36], with learning rates of 3e-5 and 5e-5 for the NYT and WebNLG datasets were explored within the range of{3e-5, 5e-5, 7e-5}. Additionally, weight decay [37] is applied with a rate of 0.01. Batch sizes of 24 for NYT and 6 for WebNLG are employed, with training conducted over 100 epochs. We utilize the cased BERT-base with 108 M parameters as the pretrained language model (PLM), setting the maximum length of the input sentence to 100 for consistency with prior research. The size of the attention head, denoted as dh, is set to 64, and the threshold $\sigma$ is established as 0.5. Parameter tuning is performed on the development set.

### 4.2  Main Results

For comparison, we employed twelve strong models as baselines. Table 1 shows that our model outperforms the other baseline methods on all the datasets. Many previous baselines had F1 scores exceeding 90% on both datasets, especially on WebNLG, which exceeded human-level performance. ORAbl achieved improvements of +0.4% and +0.5% in F1 scores on the NYT and Web databases, respectively, demonstrating the superiority of our method. The MCC metric was also better than that of the other baseline methods, which indicates that our model is still able to learn relative balance when facing imbalanced training sets. The line of $ORABL_f$ is the result of using classifier f only, indicating that each module we designed is effective.

**Table 1.** F1 scores and MCC of the model on two datasets.

| model | NYT | | WebNLG | |
|---|---|---|---|---|
| | F1. | Mcc. | F1. | Mcc. |
| GraphRel (Fu et al., 2019) | 61.9 | - | 41.1 | - |
| OrderCopyRE (Zeng et al., 2019) | 72.1 | - | 59.9 | - |
| CasRelBERT (Wei et al., 2020) | 89.6 | 93.4 | 90.1 | 91.8 |
| TPlinkerBERT (Wang et al., 2020) | 91.9 | 91.7 | 92.0 | 91.9 |
| PRGCBERT (Zheng et al., 2021a) | 92.6 | 94.0 | 92.1 | 93.0 |
| R-BPtrNetBERT (Chen et al., 2021) | 92.6 | 93.7 | 92.8 | 93.3 |
| PFN (Yan et al., 2021) | 92.4 | - | 93.6 | - |
| TDEERBERT (Li et al.,2021) | 92.5 | 86.8 | 92.4 | 93.1 |
| GRTEBERT (Ren et al.,2021) | 93.0 | 89.7 | 94.2 | 93.9 |
| EmRel (Xu et al., 2022) | 92.1 | 88.7 | 93.0 | 92.9 |
| OneRelBERT (Shang et al., 2022) | 92.8 | 89.1 | 94.4 | 88.3 |
| UniRel(Tang et al., 2023) | 93.7 | 90.8 | 94.6 | 89.7 |
| ORABL | 94.1 | 92.1 | 95.1 | 91.7 |
| $ORABL_f$ | 93.3 | 91.5 | 94.0 | 90.0 |

To further evaluate our model's performance in intricate environments, drawing inspiration from prior studies, we tested our model on three distinct subsets of the New York Times (NYT) dataset. As illustrated in Table 2, the findings underscore the efficacy of our model in navigating complex scenarios. Our model outperforms nearly all baseline models across three overlapping patterns within the normal class. Particularly noteworthy is its performance in SingleEn-tityOverlap (SEO) and EntityPairOverlap (EPO), where it achieves peak performance, showcasing ORABL's advantage in effectively managing overlapping triplets.

**Table 2.** F1 scores and MCC of the model in addressing different overlapping patterns.

| Model | Normal | SEO | EPO | SOO |
|---|---|---|---|---|
| TDEER | 90.8 | 94.1 | 94.5 | - |
| GRTE | 91.1 | 94.4 | 95.0 | - |
| OneRel | 90.6 | 95.1 | 94.8 | 90.8 |
| UniRel | 91.6 | 95.3 | 95.2 | 89.8 |
| ORABL | 92.0 | 95.6 | 95.6 | 92.1 |

To further probe into the impact of the ABL framework on model performance, we crafted a few-shot setting to scrutinize the effectiveness of the logic modules. We extracted diverse quantities of instances from every category in the training dataset (full-way) for experimental purposes. As illustrated in Table 3, our approach consistently showcases unmatched performance across virtually all scenarios, underscoring its ability to address few-shot learning challenges.

**Table 3.** F1 in fewshot.

| Model | 1 shot | 5 shot | 20 shot | 50 shot | full-shot |
|-------|--------|--------|---------|---------|-----------|
| GRTE | 58.4±5.7 | 75.2±1.9 | 86.2±0.9 | 90.3±0.4 | 93.0±0.1 |
| OneRel | 47.1±5.2 | 65.1±2.9 | 80.7±1.2 | 87.0±0.3 | 92.0±0.1 |
| UniRel | 58.2±5.8 | 75.5±2.3 | 85.6±1.0 | 90.5±0.3 | 93.5±0.1 |
| ORABL | 69.9±4.3 | 82.8±2.0 | 88.7±0.8 | 91.5±0.4 | 93.7±0.3 |

### 4.3   Analysis

We have conducted supplementary experiments aimed at evaluating the model's performance under distinct conditions. Subsequently, we meticulously analyzed the contributions and underlying mechanisms of each module towards the observed performance outcomes.

As shown in Fig. 3, as the number of shots increases, the overall performance of the model improves. In the 50-shot scenario, the model achieved 97.6% better performance than in the full-shot scenario. In terms of score composition, the precision score was always higher than the F1 score and recall score. We attribute this phenomenon to the characteristics of logic modules, which typically have higher certainty and lower triggering rates.

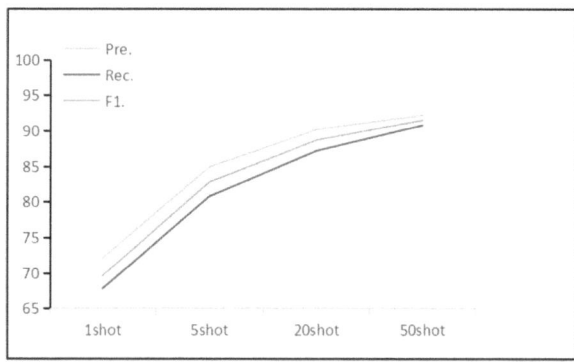

**Fig. 3.** The performance of the proposed model in different shot scenarios.

In Fig. 4, we compare the performances of the ORABL, ORABL-f (classifier only), and UniRel models in a few-shot scenario. A comparison shows that our

Abductive module can effectively improve the performance of the model under few-shot conditions. As the number of shots increases, ORABL can still maintain its advantage.

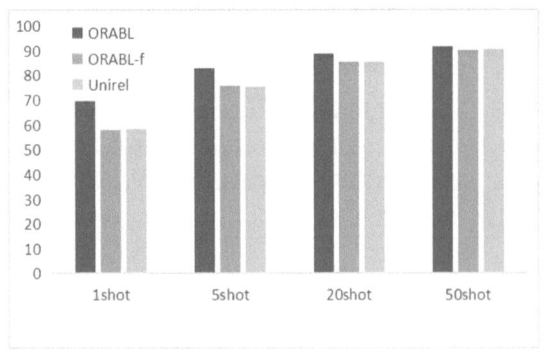

**Fig. 4.** F1 scores of different models in fewshot scenarios.

## 5    Conclusion and Future Work

In this work, we propose that ORAbL uses the ABL framework to handle the logical relations between triples and eliminate internal contradictions in the model. Through experiments, it has been proven that our designed model can effectively learn the ability to capture attention scores of entities and relations in collaboration with the ABL framework, in order to output more reliable results in complex environments. At the same time, it demonstrates certain potential in few-shot scenarios.

In the future, we will attempt to extend the ORAbl method to longer text analysis tasks in more specialized fields to obtain more prior knowledge support. Further explore the advantages of the ABL framework in few-shot scenarios. It's worth noting that in the experiments presented in Table 3, our model's F1 scores in few-shot scenarios exhibit significant variability, with fluctuations of up to 4.3 in the 1-shot setting. In future work, we will prioritize enhancing the robustness of the model to address this issue.

**Acknowledgements.** This work is supported by grant from the Natural Science Foundation of China (No. 62076046).

# References

1. Nayak, T., Majumder, N., Goyal, P., Poria, S.: Deep neural approaches to relation triplets extraction: a comprehensive survey. Cogn. Comput. **13**(5), 1215–1232 (2021)
2. Zelenko, D., Aone, C., Richardella, A.: Kernel methods for relation extraction. In: Proceedings of the ACL-02 Conference on Empirical Methods in Natural Language Processing - Volume 10, EMNLP '02, pp. 71–78, USA (2002). Association for Computational Linguistics
3. Li, Q., Ji, H.: Incremental joint extraction of entity mentions and relations. In: Toutanova, K., Wu, H., editors, Proceedings of the 52nd Annual Meeting of the Association for Computational Linguistics (Volume 1: Long Papers), pp. 402–412, Baltimore, Maryland (2014). Association for Computational Linguistics
4. Miwa, M., Bansal, M.: End-to-end relation extraction using LSTMs on sequences and tree structures. In: Erk, K., Smith, N.A., editors, Proceedings of the 54th Annual Meeting of the Association for Computational Linguistics (Volume 1: Long Papers), pp. 1105–1116, Berlin, Germany (2016). Association for Computational Linguistics
5. Wei, Z., Su, J., Wang, Y., Tian, Y., Chang, Y.: A novel cascade binary tagging framework for relational triple extraction. In: Jurafsky, D., Chai, J., Schluter, N., Tetreault, J., eds, Proceedings of the 58th Annual Meeting of the Association for Computational Linguistics, pp. 1476–1488 (2020). Association for Computational Linguistics
6. Zheng, H., et al.: PRGC: potential relation and global correspondence based joint relational triple extraction. In: Zong, C., Xia, F., Li, W., Navigli, R., eds, Proceedings of the 59th Annual Meeting of the Association for Computational Linguistics and the 11th International Joint Conference on Natural Language Processing (Volume 1: Long Papers), pp. 6225–6235 (2021). Association for Computational Linguistics
7. Wang, Y., Yu, B., Zhang, Y., Liu, T., Zhu, H., Sun, L.: TPLinker: single-stage joint extraction of entities and relations through token pair linking. In: Scott, D., Bel, N., Zong, C., eds, Proceedings of the 28th International Conference on Computational Linguistics, pp. 1572–1582, Barcelona, Spain (2020). International Committee on Computational Linguistics
8. Shang, Y.-M., Huang, H., Mao, X.: OneRel: joint entity and relation extraction with one module in one step. In: Proceedings of the AAAI Conference on Artificial Intelligence, vol. 36, issue 10, pp. 11285–11293 (2022)
9. Tang, W., et al.: UniRel: unified representation and interaction for joint relational triple extraction. arXiv preprint arXiv:2211.09039 (2022)
10. Zhou, Z.-H.: Abductive learning: towards bridging machine learning and logical reasoning. Sci. China Inf. Sci. **62**(7), 76101 (2019)
11. Yu, X., Lam, W.: Jointly identifying entities and extracting relations in encyclopedia text via a graphical model approach. In: Huang, C.-R., Jurafsky, D., eds, Coling 2010: Posters, pp. 1399–1407, Beijing, China (2010). Coling 2010 Organizing Committee
12. Miwa, M., Sasaki, Y.: Modeling joint entity and relation extraction with table representation. In: Proceedings of the 2014 Conference on Empirical Methods in Natural Language Processing (EMNLP), pp. 1858–1869 (2014)

13. Zheng, S., Wang, F., Bao, H., Hao, Y., Zhou, P., Xu, B.: Joint extraction of entities and relations based on a novel tagging scheme. In: Barzilay, R., Kan, M.-Y. eds Proceedings of the 55th Annual Meeting of the Association for Computational Linguistics (Volume 1: Long Papers), pp. 1227–1236, Vancouver, Canada (2017). Association for Computational Linguistics

14. Zeng, X., Zeng, D., He, S., Liu, K., Zhao, J.: Extracting relational facts by an end-to-end neural model with copy mechanism. In: Gurevych, I., Miyao, Y., editors, Proceedings of the 56th Annual Meeting of the Association for Computational Linguistics (Volume 1: Long Papers), pp. 506–514 Melbourne, Australia, July 2018. Association for Computational Linguistics

15. Nayak, T., Ng, H.T.: Effective modeling of encoder-decoder architecture for joint entity and relation extraction. In: Proceedings of the AAAI Conference on Artificial Intelligence, vol. 34, issue 05, pp. 8528–8535 (2020)

16. Ren, F., et al.: A novel global feature-oriented relational triple extraction model based on table filling. In: Moens, M.-F., Huang, X., Specia, L., Yih, S.W.-T., editors, Proceedings of the 2021 Conference on Empirical Methods in Natural Language Processing, pp. 2646–2656, Online and Punta Cana, Dominican Republic (2021). Association for Computational Linguistics

17. Yuan, Y., Zhou, X., Pan, S., Zhu, Q., Song, Z., Guo, L.: A relation-specific attention network for joint entity and relation extraction. In: Bessiere, C., ed, Proceedings of the Twenty-Ninth International Joint Conference on Artificial Intelligence, IJCAI-20, pp. 4054–4060. International Joint Conferences on Artificial Intelligence Organization (2020). Main track

18. Li, X., Luo, X., Dong, C., Yang, D., Luan, B., He, Z.: TDEER: an efficient translating decoding schema for joint extraction of entities and relations. In: Moens, M.-F., Huang, X., Specia, L., Yih, S.W-T., eds, Proceedings of the 2021 Conference on Empirical Methods in Natural Language Processing, pages 8055–8064, Online and Punta Cana, Dominican Republic (2021). Association for Computational Linguistics

19. Wu, H., Shi, X.: Synchronous dual network with cross-type attention for joint entity and relation extraction. In: Moens, M.-F., Huang, X., Specia, L., Yih, S.W.-T., eds, Proceedings of the 2021 Conference on Empirical Methods in Natural Language Processing, pp. 2769–2779, Online and Punta Cana, Dominican Republic (2021). Association for Computational Linguistics

20. Yan, Z., Zhang, C., Fu, J., Zhang, Q., Wei, Z.: A partition filter network for joint entity and relation extraction. In: Moens, M.-F., Huang, X., Specia, L., Yih, S.W-T., eds, Proceedings of the 2021 Conference on Empirical Methods in Natural Language Processing, pp. 185–197, Online and Punta Cana, Dominican Republic (2021). Association for Computational Linguistics

21. Xu, B., et al.: EmRel: joint representation of entities and embedded relations for multi-triple extraction. In: Carpuat, M., de Marneffe, M.-C., Ruiz, I.V.M., eds, Proceedings of the 2022 Conference of the North American Chapter of the Association for Computational Linguistics: Human Language Technologies, pp. 659–665, Seattle, United States (2022). Association for Computational Linguistics

22. Han, X., Zhao, W., Ding, N., Liu, Z., Sun, M.: PTR: prompt tuning with rules for text classification. AI Open **3**, 182–192 (2022)

23. Chen, X., et al.: KnowPrompt: knowledge-aware prompt-tuning with synergistic optimization for relation extraction. In: Proceedings of the ACM Web Conference 2022, WWW '22, pp. 2778–2788, New York, NY, USA, 2022. Association for Computing Machinery

24. Benfeng, X., Wang, Q., Lyu, Y., Zhu, Y., Mao, Z.: Entity structure within and throughout: Modeling mention dependencies for document-level relation extraction. In: Proceedings of the AAAI Conference on Artificial Intelligence, vol. 35, pp. 14149–14157 (2021)

25. Huang, Y.-X., et al.: Semi-supervised abductive learning and its application to theft judicial sentencing. In: 2020 IEEE International Conference on Data Mining (ICDM), pp. 1070–1075. IEEE (2020)

26. Lee, H., Hwang, S.J., Shin, J.: Self-supervision and self-distillation, Rethinking data augmentation (2019)

27. van den Oord, A., Li, Y., Vinyals, O.: Representation learning with contrastive predictive coding. arXiv preprint arXiv:1807.03748 (2018)

28. Lin, T.-Y., Goyal, P., Girshick, R., He, K., Dollár, P.: Focal loss for dense object detection. In: Proceedings of the IEEE International Conference on Computer Vision, pp. 2980–2988 (2017)

29. Devlin, J., Chang, M.-W., Lee, K., Toutanova, K.: BERT: pre-training of deep bidirectional transformers for language understanding. In: Burstein, J., Doran, C., Solorio, T., eds, Proceedings of the 2019 Conference of the North American Chapter of the Association for Computational Linguistics: Human Language Technologies, Volume 1 (Long and Short Papers), pp. 4171–4186, Minneapolis, Minnesota (2019). Association for Computational Linguistics

30. Vaswani, A., et al.: Attention is all you need. In: Advances in Neural Information Processing Systems, vol. 30 (2017)

31. Riedel, S., Yao, L., McCallum, A.: Modeling relations and their mentions without labeled text. In: Balcázar, J.L., Bonchi, F., Gionis, A., Sebag, M., eds, Machine Learning and Knowledge Discovery in Databases, pp. 148–163. Springer, Berlin, Heidelberg (2010)

32. Gardent, C., Shimorina, A., Narayan, S., Perez-Beltrachini, L.: Creating training corpora for NLG micro-planners. In: Barzilay, R., Kan, M.-Y., eds, Proceedings of the 55th Annual Meeting of the Association for Computational Linguistics (Volume 1: Long Papers), pp. 179–188, Vancouver, Canada (2017). Association for Computational Linguistics

33. Zhang, Y., Zhong, V., Chen, D., Angeli, G., Manning, C.D.: Position-aware attention and supervised data improve slot filling. In: Conference on Empirical Methods in Natural Language Processing (2017)

34. Han, X., et al.: FewRel: a large-scale supervised few-shot relation classification dataset with state-of-the-art evaluation. arXiv preprint arXiv:1810.10147 (2018)

35. Chicco, D., Jurman, G.: The advantages of the Matthews correlation coefficient (MCC) over f1 score and accuracy in binary classification evaluation. BMC Genomics **21**, 1–13 (2020)

36. Kingma, D.P., Ba, J.: Adam: a method for stochastic optimization. arXiv preprint arXiv:1412.6980 (2014)

37. Loshchilov, I., Hutter, F.: Decoupled weight decay regularization. arXiv preprint arXiv:1711.05101 (2017)

# Multi-task Instruction Tuning for Temporal Question Answering over Knowledge Graphs

Miao Su, Zixuan Li, Xiaolong Jin$^{(\boxtimes)}$, and Jiafeng Guo$^{(\boxtimes)}$

CAS Key Lab of Network Data Science and Technology, Institute of Computing Technology, Chinese Academy of Sciences, Beijing 100190, China
{sumiao22z,lizixuan,jinxiaolong,guojiafeng}@ict.ac.cn

**Abstract.** Answering natural language questions with temporal intent on knowledge graphs (TKGQA) has attracted rising attention in recent years. TKGQA contains several subtasks, such as implicit expression parsing, relevant facts searching, and subgraph logic reasoning, and they are composed in a pipelined manner to finish the final task. Previous work learns isolated models for each subtask, which severely restricts the knowledge sharing between related subtasks and settings, leading to underperformance. Recently, Large Language Models (LLMs) have unlocked strong multi-task capabilities from reading instructive prompts. Motivated by these, we propose a unified generation framework based on instruction tuning, called InstructTQA, for TKGQA, which unifies the key subtasks in TKGQA via text generation. Besides, to improve the time-sensitivity of LLMs, we also design several auxiliary subtasks. By capturing inter-task dependency and collaboratively learning general knowledge from different subtasks, InstructTQA demonstrates superior performance compared to the state-of-the-art one on the MultiTQ dataset, thereby substantiating its efficacy in addressing the TKGQA task.

**Keywords:** Large Language Model · Question Answering · Temporal Knowledge Graph

## 1 Introduction

In practical applications, factual knowledge tends to evolve over time [9,10, 21,24]. For instance, the host city of the Winter Olympic Games in 2018 was South Korea, while in 2022, it shifted to Beijing. There is a growing inclination towards incorporating Knowledge Graphs (KGs) with time, and these KGs are coined as TKGs. TKGs are characterized by equipping fact triplets with temporal information such as timestamps. An example representation of temporal fact in TKG is *(Beijing, hold, Winter Olympic Games, 2022)*.

To exploit the value of TKGs, recent research efforts have been devoted to answering natural language questions with temporal intent over TKG, i.e.,

© The Author(s), under exclusive license to Springer Nature Singapore Pte Ltd. 2025
X. He et al. (Eds.): CCIR 2024, LNCS 15418, pp. 94–108, 2025.
https://doi.org/10.1007/978-981-96-1710-4_8

question answering over TKG (TKGQA in short) [30]. Given a question and a background TKG, it retrieves an answer to the question from the TKG.

TKGQA can be divided into several subtasks. SubGTR [6] divided the TKGQA task into three subtasks: implicit expression parsing, relevant facts searching, and subgraph logic reasoning. MultiQA [5] refines CronKGQA by employing five modules: TKG preprocess, question preprocess, question embedding, multi-granularity time aggregation, and answer score.These existing works learn isolated models for each subtask or module. SubGTR learns a TKG embedding model and a pre-trained language model separately, while MultiQA further trains a transformer encoder on this basis. The learning of isolated models severely restricts the knowledge sharing between related tasks and settings, leading to underperformance.

Large Language Models(LLMs) [2,22,23] exhibit significant potential in achieving generalization across observed tasks through multi-task training and unified encoding [17,19,33]. It possesses the capability to generate answers for questions solely based on their internal parameter knowledge without necessitating additional fine-tuning steps or external information, thereby enabling effective performance in question-answering tasks. However, due to the possibility of incomplete, inaccurate, and outdated knowledge within LLMs [31], they often produce factually incorrect responses commonly referred to as hallucinations. Furthermore, LLMs struggle with exact temporal expressions and understanding the states and orders of multiple events [12], this further hampers the performance of LLMs on temporal questions.

In this paper, we introduce InstructTQA, a knowledge-augmented question-answering framework utilizing multi-task instruction tuning. Specifically, we reformulate the TKGQA task into subtasks framed as natural language generation tasks for the TKGQA task.

We decompose the TKGQA task into several key subtasks: entity extraction and linking, relationship generation, temporal expression generation, candidate knowledge acquisition, and answer generation. The LLM handles the main generation subtasks. To ensure the model comprehends the different tasks, we design descriptive instructions in the multi-task instruction tuning schema. Additionally, we propose auxiliary tasks to help the model capture common temporal information and enhance its understanding of temporal constraints. Specifically, we introduce temporal word generation tasks as auxiliary tasks.

We then mix these generation tasks together and conduct multitasking training on the LLM. In the reasoning stage, we perform the answer generation task to obtain the final answer.

Our main contributions are summarized as follows:

- We present a knowledge-augmented question-answering framework for TKGQA—InstructTQA, which leverages natural language instructions to guide large language models for multiple subtasks of TKGQA.
- We propose to use a retrieval augmented generation (RAG) architectural approach to improve the efficacy of LLM in answering temporal factual questions.

- Experimental results show that InstructTQA achieves state-of-the-art performance on MultiTQ and comparable results on Complex-CronQuestions.

## 2  Related Work

### 2.1  TKGQA Methods

The early TKGQA method decomposes the question into non-temporal parts and time constraints. It uses the traditional KGQA method to answer the non-temporal part and filter the candidate answers with time constraints [14]. However, this method doesn't cope with complex questions [15].

In recent years, the embedding-based TKGQA method has gradually taken over the mainstream. These methods encode entities, relations, and timestamps in TKG using a TKG embedding model and compute the similarity of question embedding and TKG embeddings to rank the candidate answers. CronKGQA [26] has first proved that TKG embeddings can be applied to the task of TKGQA. SubGTR [6] comprises three modules: implicit expression parsing, relevant facts searching, and subgraph logic reasoning. First, the question is simplified using background knowledge from the TKG to acquire explicit time constraints. Then, candidate entities are retrieved from the temporal knowledge graph using embeddings, and an initial score is computed. Finally, the answer entities are obtained by constructing a question subgraph, and time constraints are used to crop and logically reason about them. MultiQA [5] has five modules: TKG preprocess, question preprocess, question embedding, multi-granularity time aggregation, and answer score. Each module has its model parameters. This method tends to solve questions with multi-temporal granularities.

Despite achieving some success, these methods require many modules with separate parameters and lack completeness and integration. Therefore, we unified each sub-task into a generation task and proposed an end-to-end TKGQA method, InstructTQA.

### 2.2  LLM for KGQA

The combination of LLMs and knowledge graphs has attracted the attention of several researchers [20,28,35]. KAPING [1] retrieves and injects relevant knowledge directly as input prompts for LLMs. It uses a KG with symbolic knowledge in triples (subject, relation, object). These triples are verbalized by concatenating their parts. During retrieval, the verbalized triples and questions are embedded into vector space using a sentence embedding model. The similarity between these vectors is calculated to retrieve relevant triples. These triples are then concatenated with the questions to form a prompt and fed into the question-answering model to get answers. We follow the same idea as they do, where relevant knowledge is retrieved first, and then the retrieved content is used as input to the larger language model for answer generation. However, unlike KGQA, TKGQA has complex time constraints in its questions. It is still a question of whether the large language model can correctly understand the temporal words and expressions in the question and reason correctly about the answer.

Sen et al. [28] improves the retrieval method of KAPING. This work utilizes Rigel [25,29] to predict the distribution over relations for each hop and retrieve triples based on the calculated relation distribution and question entities. However, this introduces additional model training in addition to LLM.

Wu et al. [34] further transform KG knowledge into well-textualized statements most informative for KGQA and enhanced LLMs framework for solving the KGQA task. However, these methods are done with a zero-shot setting and do not provide optimal results. Recently, with the development of parameter-efficient fine-tuning methods, it has become possible to fine-tune large language models by only training a small set of parameters. In this study, we adopt Low-Rank Adaptation, or LoRA [11], to fine-tune the LLM for better performance on TKGQA.

## 3 Preliminary

In this section, we will provide a detailed formal definition of the TKGQA task.

**Temporal Knowledge Graph.** A TKG usually is denoted as $\mathcal{G} = (\mathcal{E}, \mathcal{R}, \mathcal{T}, \mathcal{F})$, where $\mathcal{E}$, $\mathcal{R}$, $\mathcal{T}$, and $\mathcal{F}$ represent the entities, relations, timestamps, and facts respectively [3]. A temporal fact $f \in \mathcal{F}$ comprises one or more entities, relations, and associated timestamps.

**Temporal Question.** A temporal question contains at least one temporal constraint or requires timestamps as its answer [13]. A temporal constraint sets a condition about a specific time point or interval that the answer must satisfy; it consists of a temporal expression and a temporal word (e.g., "in 2024"). Temporal expressions refer to time points or intervals with varying levels of granularity in natural language (e.g., "July 8th, 2024") [30]. Temporal words indicate the temporal relationships between temporal expressions and act as trigger words that impose constraints on the answers (e.g., "the same", "after", or "in the year of").

**Temporal Knowledge Graph Question Answering.** Given the temporal knowledge graph $\mathcal{G}$ and a temporal question $q$ in natural language, the TKGQA task aims to answer the $q$ using either a set of entities or timestamps from $\mathcal{G}$.

## 4 Methodology

In this section, we first introduce the overall InstructTQA process in 4.1. Then we introduce the multi-task instruction tuning schema in 4.2. Then, how the main and auxiliary tasks are mapped to the schema is introduced in Sect. 4.3 and Sect. 4.4, respectively.

### 4.1   Model Overview

**LLM Prompting with Temporal Knowledge Graphs.** We followed the approach described in KAPING to retrieve and inject relevant knowledge directly as input, referred to as "instruction," to LLMs. Our knowledge source is a TKG containing symbolic knowledge as a quadruple: *(head entity, relation, tail entity, timestamp)*. We call the set of quadruples necessary for answering the question candidate facts. To identify candidate facts, we need to extract three key pieces of information from the question: entities, relations, and temporal expressions.

**Knowledge Access** The initial step involves extracting and linking these key elements with the corresponding elements in the TKG. Subsequently, based on the question type and the linked key components, the model queries the TKG for relevant quadruples of the question.

**Knowledge Verbalization** These quadruples are then verbalized— transforming the symbolic relational knowledge into textual strings, and prepended to the input question, which is forwarded to the LLMs to generate the answer. This process enables LLMs to generate factual answers conditioned on accurate knowledge, thereby mitigating the issue of hallucinations while keeping the LLMs' parameters unchanged.

**Task Selection and Design.** We assigned LLMs to the generation of relations, temporal expressions, and answers as the main tasks in Fig. 1, while the rule-based method handled the extraction of entities and the retrieval of candidate facts. We view the generation of temporal words as auxiliary tasks in Fig. 1 to enhance the LLMs' understanding of time. Multi-task instruction tuning is conducted to train all the generation tasks.

**Main Task Selection for Generation** Initially, we employed a rule-based method to extract entities, relations, and temporal constraints from the questions. While the rule-based approach yielded good results for entity and explicit temporal constraints extraction, it struggled with relation extraction. To evaluate this, we manually constructed a dataset of 300 sample questions with corresponding relation labels. The results showed that the rule-based method achieved only 0.723 precision for relation extraction and linking, which significantly affected the subsequent retrieval of candidate facts. To address this, we leveraged LLMs to generate the relation of a question, thereby improving the precision for relation extraction and, finally, the recall of candidate facts. More details will be introduced in Sect. 4.3.

**Auxiliary Task Design** Moreover, while LLMs can answer questions, they sometimes ignore or mishandle temporal constraints. For example, in response to the question, "With which country did Swaziland last want to negotiate in 2008?", the LLM might list all countries Swaziland wanted to negotiate with in 2008 but fail to identify the last one, thus providing an incorrect answer. Approximately 30% of questions were incorrectly answered due to mishandling of time constraints. To rectify this, we guided the LLM to be more sensitive to time constraints by using it to generate temporal words and expressions.

## 4.2   Multi-task Instruction Tuning Schema

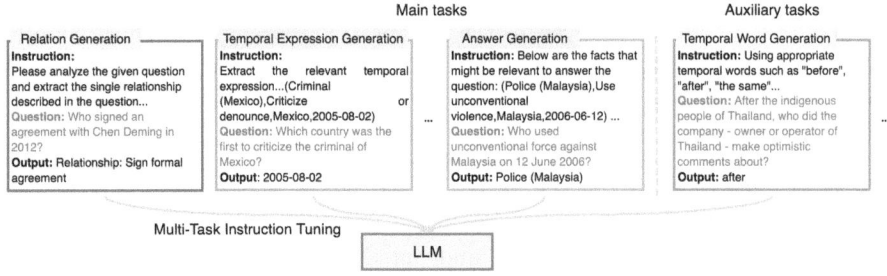

**Fig. 1.** The overview framework of InstructTQA.

Multi-task instruction tuning enables the LLMs to capture inter-task dependencies and collaboratively learn general knowledge from different subtasks. We first introduce the unified sequence-to-sequence (seq2seq) generation schema.

As shown in Fig. 1, every instance is formatted into three parts: instruction, input question, and output.

**Instruction** provides a detailed guide for a subtask, guiding the LLM to generate the target structure in natural language. We design descriptive instructions to enable the model to understand different subtasks. As shown in Fig. 1, these instructions may include information about necessary knowledge, possible output options, the format of the output structure, and any constraints or rules of the given task.

**Input question** provides the question to be answered, prefixed with "Question": to distinguish it from the instruction. It is shown in orange font in the Fig. 1.

**Output** is the target structure to be generated. It is constructed according to the specific characteristics of each subtask, such as a date in standard format or relations and entities in the TKG.

### 4.3   Main Tasks

As mentioned in Fig. 1, there are three main tasks for LLM: relation generation, temporal expression generation, and answer generation. The results of these main tasks will participate in the TKGQA task. The results of the relation generation task and the temporal expression generation task are used to retrieve the candidate facts required in the answer generation task. The answer generation task generates a final answer based on these facts.

**Relation Generation** task is to generate the relation between the entities in the question. The main objective of this task is to map the relation described in natural language in the question to the predicate in the TKG for candidate facts retrieval. The output of this task directly gives out a single relationship that is described in the question and belongs to the TKG.

**Temporal Expression Generation** task is to generate the timestamp of the temporal constraint in the question. There are two types of temporal expression; the first type is explicitly given in the question. For example, "Which country first visited the United States in 2015?" the explicit temporal expression of this question is "2015". The second type of temporal expression is implicit, where the temporal expression is given by a single event. For example, "Who commended the Military of Mali before the Armed Rebel of Mali did?" the temporal expression of this question is the timestamp of "Armed Rebel of Mali commended Military of Mali". We will retrieve the event time from TKG according to the entities and relation in the question and put the quadruple of the event in the instruction. Finally, the model is instructed to output the temporal expression in a standard date format.

**Answer Generation** task is to generate the final answer to the input question. Knowledge internalized in LLM parameters is far from enough to answer the question; we need to inject knowledge from the external TKG. A TKG consists of a set of quadruples *(head entity, relation, tail entity, timestamp)*, which include the reasoning clues and correct answers needed to answer the question; For example, for the question "After *Itthaporn Subhawong*, who was the first to *visit Cambodia?*", we additionally augment its relevant quadruples: *(Itthaporn Subhawong, Make a visit, Cambodia, 2009-09-15)* and *(Cao Duc Phat, Make a visit, Cambodia, 2009-09-23)* to the task schema. The first quadruple indicates the temporal constraint of the question is "after 2009-09-15", and the second quadruple contains the answer "Cao Duc Phat" of the question. By doing so, LLMs can generate the correct answer concerning the temporal constraint.

These candidate facts from TKG are according to the entities, relation, timestamps, and question type. They are verbalized and concatenated in the instruction; each quadruple is separated by a special token. Finally, the model is instructed to output the answers in a given format.

### 4.4   Auxiliary Tasks

To boost the LLM's understanding of temporal constraints, we further design auxiliary tasks to be trained in conjunction with the main tasks. The result of auxiliary tasks is not used in the TKGQA task procedure.

**Temporal Word Generation** task is to generate the temporal word related to the question. In our analysis of the initial wrong case, we found that the model often got temporal words like "before" and "after", "first", and "last" wrong. So, we design an auxiliary task to let the LLM generate the temporal word that describes the sequence or timing of events in the question to increase the attention of LLM to temporal words. As shown in Fig. 1, for the input question, the LLM generates the word "after" in the question.

For questions of "equal" type, there may be no temporal words in the question; we define the time granularity of these questions as their temporal word. Time granularities in the MultiTQ dataset include "year", "month", and "day". The LLM is requested to judge one from these granularities.

## 5   Experiments

### 5.1   Experimental Setup

**Datasets.** We use two datasets, MultiTQ [5] and Complex-CronQuestions [6], to demonstrate the performance of our models to the fullest. We summarize the statistics of questions in these datasets across different types in Table 1 and Table 2.

**Table 1.** Statistics of question categories in MultiTQ.

|          |              | Train   | Dev    | Test   |
|----------|--------------|---------|--------|--------|
| **Single** | Equal        | 135,890 | 18,983 | 17,311 |
|          | Before/After | 75,340  | 11,655 | 11,073 |
|          | First/Last   | 72,252  | 11,097 | 10,480 |
| **Multiple** | Equal Multi  | 16,893  | 3,213  | 3,207  |
|          | After First  | 43,305  | 6,499  | 6,266  |
|          | Before Last  | 43,107  | 6,532  | 6,247  |
| **Total** |              | 386,787 | 57,979 | 54,584 |

**Table 2.** Statistics of question categories in Complex-CronQuestions.

|              | Train  | Valid | Test  |
|--------------|--------|-------|-------|
| before_after | 6,498  | 797   | 907   |
| first_last   | 24,041 | 3,433 | 3,303 |
| time_join    | 5256   | 790   | 796   |
| **Total**    | 35,795 | 5,020 | 5,006 |

**MultiTQ** is a complex temporal question-answering dataset with multi-granularity temporal information. Questions in MultiTQ contain three temporal granularities, i.e., day, month, and year. It is constructed based on ICEWS05-15 [8]. The dataset features a few advantages, including large scale, ample relations, and multiple temporal granularity, which hence better reflects real-world scenarios.

**Complex-CronQuestions** is a complex temporal question-answering dataset based on CronQuestions, which removes all simple and pseudo-temporal questions in CronQuestions. For instance, for the question "What's the first award Carlo Taverna got?" there is only one fact related to Carlo Taverna in the TKG, which makes the temporal word "first" meaningless as a constraint. To solve this, the questions with relevant facts less than 5 are filtered by the author. This dataset consists of two parts: a KG with temporal annotations and a set of natural language questions requiring temporal reasoning. It is based on the filtered WikiData knowledge graph with 328k facts, out of which 5k are event facts.

**Baselines**

- **Pre-tained LMs:** We follow MultiQA to use BERT for MultiTQ. LM-based question embedding is generated and concatenated with the entity and time embeddings, followed by a learnable projection. The result embedding is scored against all entities and timestamps via dot-product.
- **EmbedKGQA:** This method is designed for static KGs. Timestamps are ignored during pre-training and random time embeddings are used for this model.
- **CronKGQA:** This method is designed for single temporal granularity. To deal with the MultiTQ dataset, time embeddings at the year/month granularity are drawn randomly from corresponding day embeddings.
- **MultiQA:** MultiQA improves CronKGQA by adding a multi-granularity time aggregation module to deal with different time granularities.
- **Prog-TQA:** Prog-TQA leverages the in-context learning ability of LLMs to generate question-related program drafts. Then, it aligns these drafts to TKGs and executes them to generate the answers.
- **EntityQR:** Based on EaE, EntityQR utilizes a TKG embedding-based scoring function for answer prediction.
- **TempoQR:** Similar to EntityQR, TempoQR utilizes a TKG embedding-based scoring function for answer prediction and fuses additional temporal information.
- **SubGTR:** SubGTR developed a logical reasoning module for time constraints and proposes to infer answers satisfying time constraints via temporal subgraph inference.

**Evaluation Metrics.** We report Hits@k, which measures the proportion of times the correct item appears in the top k-ranked items returned by the model, as the evaluation metrics. We use Hits@1 and Hits@10 specifically, which are the most widely used metrics for the TKGQA task. We split the output of LLM by "\n" for an answer list. If the predicted answer exactly matches the golden answer, it is considered correct.

**Implementation Details.** InstructTQA is based on LLaMA2-chat-7B [32]. We conduct supervised instruction fine-tuning based on the PEFT method LowRank Adaptation (LoRA) [11]. We set the LoRA rank and LoRA alpha parameters to 8 and 16 and turn on the LoRA target modules of q_proj, k_proj, v_proj and out_proj. We limit the sequence length to 1024 and set the batch size to 64.

During the inference phase, we use the vLLM library [16] to accelerate the LLM inference speed. We use the SentenceTransformer library for text embedding. We use the Spacy library to find entity phrases using rules describing their token lexical attributes. We use Torch for matrix acceleration when executing entity and relation linking. Experiments on LLM are conducted on 2 NVIDIA RTX3090 GPUs. Results are reported for 6 rounds of iteration on 50,000 instances for MultiTQ and 3 rounds of iteration on 143,148 fine-tuning data for Complex-CronQuestions.

**Table 3.** Overall results of InstructTQA for TKGQA task.

| | Hits@1 | | | | | Hits@10 | | | | |
|---|---|---|---|---|---|---|---|---|---|---|
| • **MultiTQ** | Overall | Question Type | | Answer Type | | Overall | Question Type | | Answer Type | |
| | | Multiple | Single | Entity | Time | | Multiple | Single | Entity | Time |
| BERT [7] | 0.083 | 0.061 | 0.092 | 0.101 | 0.040 | 0.441 | 0.392 | 0.461 | 0.531 | 0.222 |
| EmbedKGQA [27] | 0.206 | 0.134 | 0.235 | 0.290 | 0.001 | 0.459 | 0.439 | 0.467 | 0.648 | 0.001 |
| CronKGQA [26] | 0.279 | 0.134 | 0.337 | 0.328 | 0.156 | 0.608 | 0.453 | 0.671 | 0.696 | 0.392 |
| MultiQA [5] | 0.293 | 0.159 | 0.347 | 0.349 | 0.157 | 0.635 | 0.519 | 0.682 | 0.733 | 0.396 |
| Prog-TQA [4] | 0.797 | 0.750 | 0.817 | 0.790 | 0.815 | **0.934** | **0.910** | **0.944** | **0.922** | **0.963** |
| InstructTQA | **0.872** | **0.872** | **0.891** | **0.874** | **0.868** | 0.883 | 0.829 | 0.905 | 0.889 | 0.868 |
| • **Complex-CronQuestions** | Overall | Question Type | | | | Overall | Question Type | | | |
| | | before_after | first_last | time_join | | | before_after | first_last | time_join | |
| CronKGQA | 0.266 | 0.252 | 0.235 | 0.458 | | 0.767 | 0.645 | 0.777 | 0.851 | |
| EntityQR [18] | 0.425 | 0.252 | 0.235 | 0.458 | | 0.859 | 0.855 | 0.845 | 0.951 | |
| TempoQR [18] | 0.792 | 0.252 | 0.235 | 0.458 | | 0.959 | 0.923 | 0.96 | 0.990 | |
| SubGTR [6] | **0.920** | 0.847 | **0.924** | 0.977 | | **0.986** | 0.967 | **0.989** | 0.991 | |
| InstructTQA | 0.900 | **0.872** | 0.891 | 0.977 | | 0.937 | 0.914 | 0.945 | **0.991** | |

## 5.2    Results and Discussion

Table 3 reports the main result of InstructTQA on MultiTQ and Complex-CronQuestions. Overall, InstructTQA achieves an absolute improvement of 7.5% from the SOTA Hits@1 result on the MultiTQ dataset. However, the Hits@10 result is unexpectedly lower. The results demonstrate InstructTQA's capability for precisely answering temporal questions.

InstructTQA also shows comparable performance with the SOTA result of Complex-CronQuestions, especially on before_after and time_join types. Since the questions in Complex-CronQuestions have multiple relevant facts, LLM may be confused by too many facts when choosing the correct answer. Besides, the SOTA model is elaborately designed for complex questions and thus has slightly better performance. Due to time and resource constraints, we did not perform hyperparameter adjusting on this dataset.

Table 4 reports in detail the result of different categories of questions on MultiTQ. InstructTQA performs well on single-type questions, including equal, before/after, and first/last. Since the max new token length is set to 128 during the inference stage, the LLM won't generate too many tokens, so the hits@10 result is not much higher than the hits@1 results. However, the extremely high hits@1 results illustrate the accuracy of InstructTQA in answering temporal questions.

**Table 4.** Results for Hits@1 and Hits@10 of MultiTQ.

| Category | Hits@1 | Hits@10 |
|---|---|---|
| After First | 0.854 | 0.854 |
| Before/After | 0.910 | 0.952 |
| Before Last | 0.793 | 0.793 |
| Equal | 0.882 | 0.887 |
| Equal Multi | 0.833 | 0.850 |
| First/Last | 0.884 | 0.884 |
| **Overall** | 0.872 | 0.883 |

### 5.3   Error Analysis

We analyze the errors in the test set of MultiTQ. As shown in Fig. 2, most of the errors come from missing answers in the knowledge (i.e. retrieved candidate facts). In this case, the LLM can not answer the questions based on the TKG facts; it can sometimes answer questions correctly based on its internalized knowledge, according to our observations. The reason for missing answers may be some mistake from entity extraction and entity linking or from the candidate retrieval process.

We further analyze the reason why the candidate facts contain answers, but the model cannot provide the correct answer, i.e. answer wrongly in Fig. 2. The most common situation is that the prompt is too long for LLM's as an input. It accounts for 53.5% of this case. This is followed by ignoring or mistaking temporal constraints and confusing head and tail entities, which account for 15.5% and 14.5%, respectively. The rest of the errors come from answering the entities in the question or lacking the knowledge needed to reason, or the LLM simply make up some answers.

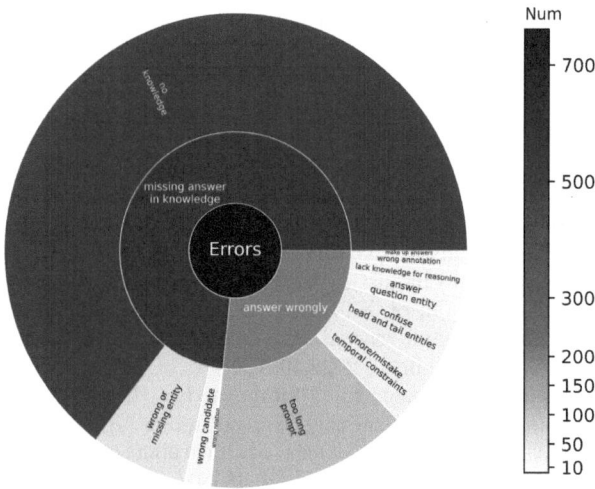

**Fig. 2.** Overview of the causes of error on MultiTQ.

## 6    Conclusions

In this paper, we proposed a knowledge-augmented question-answering framework based on multi-task instruction tuning named InstructTQA. We reformulated TKGQA tasks into subtasks of natural language generation. We divide the TKGQA task into several main tasks, including relation generation, temporal expression generation, and answer generation. We design descriptive instructions to enable the model to understand different tasks and generate the target structure in the form of natural language. We further propose auxiliary tasks, which deepen the understanding of time sequence, event orders, and time granularity through our well-designed temporal word generation task. These generation tasks are handed over to the LLM and trained under our multi-task instruction tuning schema.

We conducted experiments on datasets that presented challenges in terms of time granularity and complexity. The results showed that InstructTQA achieves state-of-the-art results on MultiTQ and comparable results on Complex-CronQuestions.

## References

1. Baek, J., Aji, A.F., Saffari, A.: Knowledge-augmented language model prompting for zero-shot knowledge graph question answering (2023). https://doi.org/10. 48550/arXiv.2306.04136
2. Brown, T., et al.: Language models are few shot learners. In: Advances in Neural Information Processing Systems, vol. 33, pp. 1877–1901. Curran Associates, Inc. (2020)

3. Cai, B., Xiang, Y., Gao, L., Zhang, H., Li, Y., Li, J.: Temporal knowledge graph completion: a survey. In: Proceedings of the Thirty-Second International Joint Conference on Artificial Intelligence, pp. 6545–6553 (2023). https://doi.org/10.24963/ijcai.2023/734

4. Chen, Z., et al.: Self improvement programming for temporal knowledge graph question answering (2024). https://doi.org/10.48550/arXiv.2404.01720

5. Chen, Z., Liao, J., Zhao, X.: Multi-granularity temporal question answering over knowledge graphs. In: Proceedings of the 61st Annual Meeting of the Association for Computational Linguistics (Volume 1: Long Papers), pp. 11378–11392. Association for Computational Linguistics, Toronto, Canada (2023)

6. Chen, Z., Zhao, X., Liao, J., Li, X., Kanoulas, E.: Temporal knowledge graph question answering via subgraph reasoning. Knowl.-Based Syst. **251**, 109134 (2022). https://doi.org/10.1016/j.knosys.2022.109134

7. Devlin, J., Chang, M.W., Lee, K., Toutanova, K.: BERT: pre-training of deep bidirectional transformers for language understanding (2018)

8. García-Durán, A., Dumančić, S., Niepert, M.: Learning sequence encoders for temporal knowledge (2018). https://doi.org/10.48550/arXiv.1809.03202

9. Gottschalk, S., Demidova, E.: EventKG: a multilingual event-centric temporal knowledge graph (2018). https://doi.org/10.48550/arXiv.1804.04526

10. Hoffart, J., et al.: Robust disambiguation of named entities in text. In: Barzilay, R., Johnson, M. (eds.) Proceedings of the 2011 Conference on Empirical Methods in Natural Language Processing, pp. 782–792. Association for Computational Linguistics, Edinburgh, Scotland, UK (2011)

11. Hu, E.J., et al.: LoRA: low-rank adaptation of large language models (2021). https://doi.org/10.48550/arXiv.2106.09685

12. Jain, R., Sojitra, D., Acharya, A., Saha, S., Jatowt, A., Dandapat, S.: Do language models have a common sense regarding time? Revisiting temporal commonsense reasoning in the era of large language models. In: Bouamor, H., Pino, J., Bali, K. (eds.) Proceedings of the 2023 Conference on Empirical Methods in Natural Language Processing, pp. 6750–6774. Association for Computational Linguistics, Singapore (2023). https://doi.org/10.18653/v1/2023.emnlp-main.418

13. Jia, Z., Abujabal, A., Saha Roy, R., Strötgen, J., Weikum, G.: TempQuestions: a benchmark for temporal question answering. In: Companion Proceedings of the The Web Conference 2018, pp. 1057–1062. WWW '18, International World Wide Web Conferences Steering Committee, Republic and Canton of Geneva, CHE (2018). https://doi.org/10.1145/3184558.3191536

14. Jia, Z., Abujabal, A., Saha Roy, R., Strötgen, J., Weikum, G.: TEQUILA: temporal question answering over knowledge bases. In: Proceedings of the 27th ACM International Conference on Information and Knowledge Management, pp. 1807–1810. ACM, Torino Italy (2018). https://doi.org/10.1145/3269206.3269247

15. Jia, Z., Pramanik, S., Roy, R.S., Weikum, G.: Complex temporal question answering on knowledge graphs. In: Proceedings of the 30th ACM International Conference on Information & Knowledge Management, pp. 792–802 (2021). https://doi.org/10.1145/3459637.3482416

16. Kwon, W., et al.: Efficient memory management for large language model serving with paged attention (2023). https://doi.org/10.48550/arXiv.2309.06180

17. Longpre, S., et al.: The Flan Collection: designing data and methods for effective instruction tuning (2023). https://doi.org/10.48550/arXiv.2301.13688

18. Mavromatis, C., et al.: TempoQR: temporal question reasoning over knowledge graphs. In: Proceedings of the AAAI Conference on Artificial Intelligence, vol. 36, issue (5), pp. 5825–5833 (2022). https://doi.org/10.1609/aaai.v36i5.20526

19. Mishra, S., Khashabi, D., Baral, C., Hajishirzi, H.: Cross-Task Generalization via Natural Language Crowdsourcing Instructions. In: Proceedings of the 60th Annual Meeting of the Association for Computational Linguistics (Volume 1: Long Papers), pp. 3470-3487. Association for Computational Linguistics, Dublin, Ireland (2022). https://doi.org/10.18653/v1/2022.acl-long.244

20. Nie, Z., Zhang, R., Wang, Z., Liu, X.: Code-style in-context learning knowledge-based (2023). https://doi.org/10.48550/arXiv.2309.04695

21. Nonaka, I., Toyama, R., Konno, N.: SECI, Ba and Leadership: a unified model of dynamic knowledge creation. Long Range Plan. **33**, 5–34 (2000). https://doi.org/10.1016/S0024-6301(99)00115-6

22. OpenAI, Achiam, J., et al.: GPT-4 technical report (2024). https://doi.org/10.48550/arXiv.2303.08774

23. Ouyang, L., et al.: Training language models to follow instructions with human feedback (2022). https://doi.org/10.48550/arXiv.2203.02155

24. Roddick, J.F., Spiliopoulou, M.: A survey of temporal knowledge discovery paradigms and methods. IEEE Trans. Knowl. Data Eng. **14**(4), 750–767 (2002). https://doi.org/10.1109/TKDE.2002.1019212

25. Saffari, A., Oliya, A., Sen, P., Ayoola, T.: End-to-end entity resolution and question answering using differentiable knowledge graphs. In: Moens, M.F., Huang, X., Specia, L., Yih, S.W.t. (eds.) Proceedings of the 2021 Conference on Empirical Methods in Natural Language Processing, pp. 4193–4200. Association for Computational Linguistics, Online and Punta Cana, Dominican Republic (2021). https://doi.org/10.18653/v1/2021.emnlp-main.345

26. Saxena, A., Chakrabarti, S., Talukdar, P.: Question answering over temporal knowledge graphs (2021)

27. Saxena, A., Tripathi, A., Talukdar, P.: Improving multi-hop question answering over knowledge graphs using knowledge base embeddings. In: Jurafsky, D., Chai, J., Schluter, N., Tetreault, J. (eds.) Proceedings of the 58th Annual Meeting of the Association for Computational Linguistics, pp. 4498–4507. Association for Computational Linguistics, Online (2020). https://doi.org/10.18653/v1/2020.aclmain.412

28. Sen, P., Mavadia, S., Saffari, A.: Knowledge graph-augmented language models for complex question answering. In: Dalvi Mishra, B., Durrett, G., Jansen, P., Neves Ribeiro, D., Wei, J. (eds.) Proceedings of the 1st Workshop on Natural Language Reasoning and Structured Explanations (NLRSE), pp. 1–8. Association for Computational Linguistics, Toronto, Canada (2023). https://doi.org/10.18653/v1/2023.nlrse-1.1

29. . Sen, P., Oliya, A., Saffari, A.: Expanding end-to-end question answering on differentiable knowledge graphs with intersection. In: Moens, M.F., Huang, X., Specia, L., Yih, S.W.t. (eds.) Proceedings of the 2021 Conference on Empirical Methods in Natural Language Processing, pp. 8805–8812. Association for Computational Linguistics, Online and Punta Cana, Dominican Republic (2021). https://doi.org/10.18653/v1/2021.emnlp-main.694

30. Su, M., Li, Z., Chen, Z., Bai, L., Jin, X., Guo, J.: Temporal knowledge graph question answering: a survey (2024). https://doi.org/10.48550/arXiv.2406.14191

31. Sun, K., Xu, Y.E., Zha, H., Liu, Y., Dong, X.L.: Head-to-Tail: how knowledgeable are large language models (LLMs)? A.K.A. Will LLMs Replace Knowledge Graphs? (2024). https://doi.org/10.48550/arXiv.2308.10168

32. Touvron, H., et al.: Llama 2: open foundation and fine-tuned chat models (2023). https://doi.org/10.48550/arXiv.2307.09288

33. Wang, Y., et al.: Super-NaturalInstructions: generalization via declarative instructions on 1600+ NLP tasks (2022). https://doi.org/10.48550/arXiv.2204.07705
34. Wu, Y., et al.: Retrieve-RewriteAnswer: a KG-to-Text enhanced LLMs framework for knowledge graph question answering (2023). https://doi.org/10.48550/arXiv.2309.11206
35. Yang, L., et al.: Give Us the Facts: Enhancing Large Language Models with Knowledge Graphs for Fact-aware Language Modeling (2024). https://doi.org/10.48550/arXiv.2306.11489

# On the Capacity of Citation Generation by Large Language Models

Haosheng Qian, Yixing Fan[✉], Ruqing Zhang, and Jiafeng Guo

CAS Key Lab of Network Data Science and Technology, Institute of Computing Technology, Chinese Academy of Sciences, Beijing 100190, China
{qianhaosheng22s,fanyixing,zhangruqing,guojiafeng}@ict.ac.cn

**Abstract.** Retrieval-augmented generation (RAG) appears as a promising method to alleviate the "hallucination" problem in large language models (LLMs), since it can incorporate external traceable resources for response generation. The essence of RAG in combating the hallucination issue lies in accurately attributing claims in responses to the corresponding retrieved documents. However, most of existing works focus on improving the quality of generated responses from the LLM, while largely overlooked its ability to attribute sources accurately. In this study, we conduct a systematic analysis about the capabilities of LLMs in generating citations within response generation, and further introduce a novel method to enhance their citation generation abilities. Specifically, we evaluate both the correctness and citation quality for seven widely-used LLMs on two benchmark datasets. Meanwhile, we introduce new citation evaluation metrics to eliminate the over-penalization of unnecessary and excessive citations in existing metrics. Furthermore, we propose a Generate-then-Refine method that completes relevant citations and removes irrelevant ones without altering the response text. The results on WebGLM-QA, ASQA and ELI5 datasets show that our method substantially improves the quality of citations in responses generated by LLMs.

**Keywords:** Large Language Model · Retrieval-Augemented Generation · Citation Generation

## 1 Introduction

Recently, large language models (LLMs) [1] demonstrate outstanding performance in various natural language processing tasks, showing remarkable generative capabilities for complex questions [2]. However, LLMs also face the well-known "hallucination" issue as they tend to produce fabricated content for unknown questions, which largely hinders the practical usage in risk-aware applications, such as medical or legal consultants. To this end, RAG appears as a promising method to incorporate real-time and factual knowledge for response generation [3,4].

© The Author(s), under exclusive license to Springer Nature Singapore Pte Ltd. 2025
X. He et al. (Eds.): CCIR 2024, LNCS 15418, pp. 109–123, 2025.
https://doi.org/10.1007/978-981-96-1710-4_9

While RAG could enhance the LLMs in leveraging external resources through in-context learning, it is crucial to acknowledge that the core is to provide citations for any generated statements in responses. However, recent advances in RAG mainly focused on building complex architectures to improve the quality of retrieved content [5]. For example, FLARE [6] retrieves information iteratively and actively by monitoring the confidence of generated tokens. ITER-RETGEN [7] also enhances retrieval by using generated content, achieving an iterative retrieval-generation flow. More recently, the attribution for response has attracted lots of attention in both academia and industry. For example, Gao et al. [8] propose ALCE, the first benchmark for automatic LLMs' citation evaluation. Besides, the Bing Chat[1] and perplexity[2] have already implemented the citation generation in their online systems.

In existing works, there are generally two types of methods to provide citations for responses: the pre-hoc citation and the post-hoc citation [9]. The pre-hoc method treats citations as regular tokens and generates them directly during the inference process of LLMs, which places high demands on the capabilities of the LLMs [10]. In contrast, the post-hoc method firstly generate a response without citations, and then matches the content of the response with references to determine whether citations need to be added [8].

In fact, the pre-hoc method for citation generation often results in better consistency between the responses and the references, as it fully leverages the LLMs' excellent natural language understanding capabilities. Nonetheless, generating accurate citations is still a significant challenge for LLMs. This task demands that LLMs analyze multiple references, provide a coherent and comprehensive response, and determine precisely when to incorporate citations.

In this study, we systematically analyze the latest LLMs' abilities in generating citations within responses generation and introduce a novel method to enhance citation quality. We use two basic methods-few-shot and fine-tuning-to guide LLMs in generating responses with citations. Then, we evaluate the correctness of responses and the quality of citations across three long-form question answer (LFQA) datasets in two benchmarks. For evaluation, we found that metrics in ALCE [8] excessively penalize responses that either don't require citations or include excessive citations. Therefore, we introduce more comprehensive metrics to evaluate the citation quality in responses. We exclude statements that don't require citations in responses for citation recall and redefine the concept of "relevant" for citation precision.

Moreover, we integrate pre-hoc and post-hoc methods to introduce Generate-then-Refine approach. This approach adds relevant citations that were not initially generated in the response and removes irrelevant citations that were included, thereby improving citation quality without altering the response text itself. We conduct experiments on three LFQA datasets: WebGLM-QA [11], ASQA [12], and ELI5 [13]. The experiment results demonstrate that our proposed method significantly improves the citation quality.

---

[1] https://www.bing.com/new.

[2] https://www.perplexity.ai.

In summary, our contributions are threefold: (1) we analyze the latest LLMs' ability to generate citations; (2) we introduce more comprehensive metrics for evaluating citation quality; (3) we propose the Generate-then-Refine approach, which substantially enhances the citation quality in responses.

## 2   Related Work

In this section, we review existing relevant work from two perspectives: citation generation and citation evaluation. Some researchers also refer to the process of associating responses with their corresponding supporting references as "attribution", and we include these works as well.

**Citation Generation.** Recently, a host of works in the RAG field have required LLMs to provide citations while generating responses. Nakano et al. [10] presented WebGPT, which fine-tunes GPT-3 to answer long-form questions based on a web browsing environment. This is one of the earliest works enabling LLMs to generate responses with citations. Menick et al. [14] used reinforcement learning from human preferences to train language models that generate responses while also citing specific evidence to support their claims. Qian et al. [15] introduced ReGen framework, which enhances the generation factualness and supports the generation of responses with citations. Liu et al. [11] presented WebGLM, employing a rule-based approach to match responses and references for filtering high-quality training data containing citations, and fine-tunes LLMs to learn incorporating citations into answers. Qin et al. [16] presented WebCPM, a fine-tuned language model that imitates human web search behavior, treating "Quote" as an action to extract content from current web page for supporting evidence during response generation. Gao et al. [8] used a few-shot method to guide LLMs in generating citations and also provide a post-hoc cite option to add citations into the responses. Sun et al. [17] introduced an approach named VTG, incorporating evolving memory and self-reflection, supporting evidence verification and retrieval. This approach aids models in rethinking and reflecting on the relationship between claims and citations. Huang et al. [18] proposed a training framework using fine-grained rewards to teach LLMs to generate highly supportive and relevant citations.

**Citation Evaluation.** It is crucial to quantitatively evaluate the quality of citations in responses once models are capable of generating citations. Rashkin et al. [19] proposed a manual evaluation framework named AIS for measuring whether model-generated statements are supported by underlying sources. Based on this, Gao et al. [20] introduced an automated metric AutoAIS, which approximates human AIS judgments using an NLI model. Bohnet et al. [21] subsequently defined a reproducible evaluation framework for Attributed QA, using human annotations as a gold standard and employing AutoAIS as an automatic evaluation metric. Liu et al. [22] manually evaluated the citations included in popular generative search engines from the perspectives of comprehensiveness and accuracy. Liu et al. [11] manually evaluated the relationships between answers generated by LLMs and their corresponding references for citation accuracy. Yue et al. [23] defined different types of attribution errors and employed two approaches, prompting LLMs and fine-tuning smaller LMs, for automatic evaluation of attribution. Gao et al. [8] proposed the first benchmark for automatic LLMs' citation evaluation—ALCE, which defines citation recall and citation precision metrics to measure citation quality, and uses an NLI model to determine whether the responses are supported by cited references. Kamalloo et al. [24] established an attribution dataset, where LLMs generate answers with citations initially, which are then annotated by human based on informativeness and attributability. Hu et al. [25] defined more fine-grained attribution categories and proposed an automatic manner for generating benchmarks of Attributed QA using knowledge graphs.

## 3    Analysis of Citation Generation by Large Language Models

In this section, we conduct a comprehensive evaluation and analysis of the latest LLMs' ability to generate citations. We employ two basic methods, few-shot and fine-tuning, to guide LLMs in generating responses with citations.

### 3.1    Datasets

We select three LFQA datasets for our experiments: (1) WebGLM-QA [11], consisting of 43,579 data samples for the train split, 1,000 for the validation split, and 400 for the test split. Each data sample contains a question, an answer and a set of references. The answers and accompanying citations in the dataset were generated by GPT-3 [1] through in-context learning. Liu et al. [11] applied a series of rules to filter the dataset. They used ROUGE-1 [26] to measure the similarity between answer segments and their corresponding references to remove irrelevant citations which were labeled inaccurately. Additionally, they also implemented rule-based filtering to alleviate issues such as hallucination, few citations, and low-quality citations. (2) ASQA [12] is a factoid QA dataset where the questions often contain ambiguities, resulting in multiple answers based on different interpretations. Responses to these ambiguous questions should synthesize factual information from multiple sources to form the final answer. ALCE benchmark [8] randomly selected 948 samples from original ASQA dataset and added retrieved passages to construct a test set. (3) ELI5 [13] is also an LFQA dataset, with questions collected from the Reddit forum "Explain Like I'm Five", primarily consisting of "How" and "Why" questions. For these types of questions, good answers are often quite detailed and cannot be adequately addressed with brief responses or by simply extracting words or phrases from the contexts. In a similar manner, ALCE benchmark [8] selected 1000 samples to construct a test set.

## 3.2   Evaluation

We evaluate responses for both their correctness and citation quality. Despite our main focus is not on correctness, it remains a crucial aspect of evaluation. We use well-established metrics BLEU-4 [27] and ROUGE-L [26] to measure correctness.

For citation quality evaluation, we initially adopt two metrics defined in ALCE [8]: citation recall and citation precision.

**Citation Recall.** Before evaluation, each response is segmented into several statements $\{s_1, s_2, ...\}$. Citation recall is computed on a per-statement basis, where each statement $s_i$ receives a binary recall score. The citation recall is the average of recall scores across all statements in the entire dataset. For each statement $s_i$, recall score is 1 if and only if $s_i$ contains at least one citation and $\phi(concat(C_i), s_i) = 1$, where $\phi(premise, hypothesis)$ is the NLI model that outputs 1 if the premise entails the hypothesis, and 0 otherwise [8]. And $concat(C_i)$ denotes the concatenation of all passages cited by $s_i$.

This calculation method is too strict for some responses. Upon reviewing the experimental results, we found that not all statements necessarily require citations. Statements that are commonsense such as "Humans can walk but cannot fly" or transitional statements like "Next, I will answer the question from the following aspects" don't need any citations. The above metric may lead to underestimated evaluations for certain responses.

Based on this issue, we have defined a more lenient metric. If a statement $s_i$ doesn't have any citation and $\phi(concat(C_{all}), s_i) = 0$, then it will not require computation of its recall score and will not be included in the final average. And $concat(C_{all})$ denotes the concatenation of all retrieved passages that serve as context prompts to the LLMs in current sample.

**Citation Precision.** Similar to citation recall, before calculating citation precision, each response is also segmented into statements. However, citation precision is calculated on a per-citation basis, where each citation $c_{ij}$ receives a binary precision score. The citation precision is the average of precision scores across all citations in the entire dataset. Citation precision focuses on whether each citation $c_{ij}$ is relevant to the statement $s_i$. A citation $c_{ij}$ is "irrelevant" if $c_{ij}$ itself cannot support $s_i$ and does not affect the rest of the citations to support $s_i$ [8].

This definition of "relevant" may overly penalize answers that have excessive citations. If two references contain the same information and happen to be cited together in a statement, the above method may misjudge both citations as irrelevant, even though they both contribute to supporting the statement.

Due to this issue, we have redefined "relevant". For a citation $c_{ij}$, if it can support statement $s_i$ independently or if it can support statement $s_i$ after combining with a subset of remaining citations $C_i \backslash \{c_{ij}\}$ that cannot support statement $s_i$, we consider this citation $c_{ij}$ is relevant. Formally, $c_{ij}$ is "relevant" if either of the following two conditions is satisfied:

$$(a) \quad \phi(c_{ij}, s_i) = 1,$$
$$(b) \quad \exists \, C_i' \subset C_i \backslash \{c_{ij}\}, \;\; \phi(\text{concat}(\{c_{ij}\} \cup C_i'), s_i) = 1. \tag{1}$$

Unlike the conciseness pursued by citation precision in ALCE, our redefined citation precision allows LLMs to generate comprehensive citations in responses. However, whether conciseness or comprehensiveness is preferable depends on specific scenarios, making it difficult to conclusively determine which evaluation approach is better. Therefore, we report the results of both metrics in our subsequent experiments.

### 3.3 Implementation Details

We conduct experiments on seven representative latest LLMs, including GPT -3.5-turbo-0125 [28], Llama-2-7b-chat [29], Llama-2-13b-chat [29], Mistral-7B-Instruct-v0.2 [30], Meta-Llama-3-8B-Instruct [31], glm-4-9b-chat [32], and Qwen2-7B-Instruct [33]. We fine-tune LLMs on the train split of WebGLM-QA and evaluate them on the test split of WebGLM-QA, as well as on the oracle versions of ASQA and ELI5 in ALCE [8,11–13]. And we use t5_xxl_true_nli_mixture [34] as the NLI model when evaluating metrics related to citation.

In few-shot experiments, we provide two examples for each input. In the fine-tuning experiments, we use LoRA [35] method to fine-tune six open-source LLMs. To facilitate reproducibility of results and avoid bias introduced by sampling during decoding, we employ greedy decoding for all open-source models.

### 3.4 Results

The main results are summarized in Tables 1, 2, 3, and we have the key observations as follows.

First, early open-source models lack the ability to generate citations. Earlier released LLMs like Llama-2 series, whether with 7B or 13B parameters, perform poorly in few-shot experiments across all three datasets. In contrast, Llama-3 models developed by the same team as Llama-2 appear to have significantly better citation generation capabilities. In few-shot experiments on ASQA and ELI5, Llama-3 even surpassed GPT-3.5-turbo by a wide margin. Additionally, other more recently released LLMs have also shown a nearly satisfactory ability in few-shot settings. One possible reason is that as LLMs are increasingly used in RAG tasks, model developers have started to focus on attribution capabilities of LLMs and have conducted additional training on related tasks for them.

Second, LLMs can significantly benefit from fine-tuning to enhance their citation generation capabilities. After fine-tuning, all open-source models demonstrate substantial improvements on WebGLM-QA, both in response correctness and citation quality, compared to their few-shot results. The previously underperforming Llama-2 series models reached performance levels comparable to other models through fine-tuning. Notably, Llama-2-13b even surpassed GPT-3.5-turbo in citation generation capabilities.

**Table 1.** Experiments on WebGLM-QA.

| Models | Correctness | | Citation (ALCE) | | | Citation (Ours) | | |
|---|---|---|---|---|---|---|---|---|
| | BLEU-4 | ROUGE-L | Recall | Precision | F1 | Recall | Precision | F1 |
| **Few-Shot** | | | | | | | | |
| gpt-3.5-turbo | 57.44 | 41.41 | **74.21** | **74.91** | **74.56** | **78.31** | **78.18** | **78.24** |
| llama2-7b | 37.27 | 40.24 | 27.85 | 54.24 | 36.80 | 28.95 | 57.33 | 38.47 |
| llama2-13b | 40.13 | 40.48 | 30.08 | 50.94 | 37.82 | 31.78 | 55.63 | 40.45 |
| mistral-7b | 65.51 | 47.21 | 73.03 | 66.39 | 69.55 | 74.31 | 62.84 | 68.10 |
| llama3-8b | 54.38 | 45.32 | 72.88 | 72.99 | 72.93 | 74.58 | 69.25 | 71.82 |
| glm4-9b | 50.38 | 44.44 | 60.98 | 73.94 | 66.84 | 62.11 | 72.09 | 66.73 |
| qwen2-7b | 50.61 | 39.86 | 63.07 | 66.55 | 64.76 | 63.56 | 65.52 | 64.53 |
| **Fine-Tuning** | | | | | | | | |
| llama2-7b | 69.61 | 55.59 | 78.96 | 79.84 | 79.40 | 79.17 | 75.94 | 77.52 |
| llama2-13b | 71.98 | 57.32 | 79.56 | 82.92 | 81.21 | 80.21 | **78.21** | **79.20** |
| mistral-7b | 70.02 | 56.31 | 79.52 | 81.52 | 80.51 | 80.13 | 77.39 | 78.74 |
| llama3-8b | 70.73 | 57.25 | **80.00** | 82.98 | **81.46** | **80.47** | 77.80 | 79.11 |
| glm4-9b | 71.31 | 57.37 | 79.07 | **83.05** | 81.01 | 80.00 | 77.94 | 78.96 |
| qwen2-7b | 70.88 | 55.07 | 77.51 | 80.46 | 78.96 | 78.48 | 72.90 | 75.59 |

Third, fine-tuned LLMs don't generalize well. The models fine-tuned on WebGLM-QA demonstrate significantly better performance on its test set compared to the original models' few-shot results. However, their results on ASQA are quite mediocre, even falling short of their few-shot results. Only the Llama-2 series models, which originally lacked attribution capabilities, showed improved performance across all three datasets after fine-tuning. This illustrates that even with similar task formats, model performance can vary greatly due to changes in data distribution.

Furthermore, GPT-3.5-turbo demonstrates notably strong attribution capabilities in few-shot experiment on WebGLM-QA compared to open-source models. However, its performance is less impressive on ASQA and ELI5. This discrepancy might be due to model's heightened sensitivity to the examples provided in few-shot method. We use identical examples as context across the three datasets in our experiments. For the model, there may be significant differences between these examples and the actual samples being evaluated.

## 4    Generate-then-Refine

Based on the previous experiments and analysis, we have identified there is still considerable room for improvement in the quality of citations within the responses. Inspired by post-hoc methods [9], we propose a Generate-then-Refine approach aimed at improving the citation quality without altering the response

**Table 2.** Experiments on ASQA.

| Models | Correctness | | Citation (ALCE) | | | Citation (Ours) | | |
|---|---|---|---|---|---|---|---|---|
| | BLEU-4 | ROUGE-L | Recall | Precision | F1 | Recall | Precision | F1 |
| **Few-Shot** | | | | | | | | |
| gpt-3.5-turbo | 28.56 | 27.18 | 52.56 | 51.87 | 52.21 | 53.99 | **67.57** | 60.02 |
| llama2-7b | 28.88 | 27.35 | 19.75 | 33.47 | 24.84 | 21.39 | 41.85 | 28.31 |
| llama2-13b | 34.34 | 28.17 | 22.25 | 36.91 | 27.76 | 23.62 | 42.81 | 30.44 |
| mistral-7b | 45.31 | 30.33 | 57.05 | 57.77 | 57.41 | 60.14 | 58.65 | 59.39 |
| llama3-8b | 23.62 | 30.57 | **67.38** | **64.77** | **66.05** | **68.53** | 67.47 | **68.00** |
| glm4-9b | 37.45 | 31.34 | 58.50 | 61.84 | 60.12 | 59.85 | 65.36 | 62.48 |
| qwen2-7b | 20.75 | 29.13 | 56.75 | 57.60 | 57.17 | 57.27 | 62.96 | 59.98 |
| **Fine-Tuning** | | | | | | | | |
| llama2-7b | 34.92 | 30.69 | 60.02 | 46.72 | 52.54 | 61.61 | 51.38 | 56.03 |
| llama2-13b | 34.37 | 30.94 | 60.35 | 45.26 | 51.73 | 62.93 | 50.71 | 56.16 |
| mistral-7b | 42.97 | 31.59 | 60.08 | 49.19 | 54.09 | 62.01 | 52.80 | 57.04 |
| llama3-8b | 40.07 | 31.75 | **63.62** | **49.58** | **55.73** | **64.02** | **52.93** | **57.95** |
| glm4-9b | 42.30 | 31.90 | 61.64 | 41.77 | 49.80 | 62.03 | 44.02 | 51.50 |
| qwen2-7b | 39.54 | 31.04 | 58.26 | 44.00 | 50.14 | 60.56 | 46.60 | 52.67 |

**Table 3.** Experiments on ELI5.

| Models | Correctness | | Citation (ALCE) | | | Citation (Ours) | | |
|---|---|---|---|---|---|---|---|---|
| | BLEU-4 | ROUGE-L | Recall | Precision | F1 | Recall | Precision | F1 |
| **Few-Shot** | | | | | | | | |
| gpt-3.5-turbo | 29.06 | 15.01 | 23.33 | 24.87 | 24.08 | 25.01 | 47.24 | 32.71 |
| llama2-7b | 22.06 | 15.78 | 15.82 | 39.13 | 22.53 | 16.76 | 42.14 | 23.98 |
| llama2-13b | 23.93 | 16.49 | 15.40 | 32.88 | 20.98 | 16.69 | 37.66 | 23.13 |
| mistral-7b | 33.50 | 16.86 | **43.73** | 40.85 | 42.24 | 45.03 | 45.95 | 45.49 |
| llama3-8b | 27.38 | 16.75 | 42.98 | **46.48** | **44.66** | **45.09** | 52.88 | **48.68** |
| glm4-9b | 26.08 | 16.61 | 29.59 | 44.54 | 35.56 | 30.95 | 48.89 | 37.90 |
| qwen2-7b | 27.90 | 15.32 | 35.40 | 41.70 | 38.29 | 35.96 | 46.14 | 40.42 |
| **Fine-Tuning** | | | | | | | | |
| llama2-7b | 31.69 | 17.53 | **49.44** | 51.66 | 50.53 | **49.93** | 55.13 | 52.40 |
| llama2-13b | 31.54 | 17.48 | 47.01 | 51.00 | 48.92 | 47.88 | 56.46 | 51.82 |
| mistral-7b | 30.60 | 17.57 | 48.92 | **54.46** | **51.54** | 49.82 | **57.42** | **53.35** |
| llama3-8b | 32.30 | 17.66 | 48.71 | 52.62 | 50.59 | 49.35 | 57.15 | 52.96 |
| glm4-9b | 31.64 | 17.61 | 48.69 | 52.70 | 50.62 | 49.84 | 55.07 | 52.32 |
| qwen2-7b | 32.40 | 17.06 | 44.81 | 52.62 | 48.40 | 45.94 | 54.41 | 49.82 |

text. Previous post-hoc methods heavily relied on rule-based matching such as text overlap, which is ineffective for semantic matching. Leveraging the powerful natural language understanding capabilities of LLMs, we aim to fine-tune the LLM to become a robust refiner in our approach.

### 4.1    Methods

We aim for the refiner to have three capabilities: (1) keep relevant citations within the response; (2) add necessary citations that are missing; (3) remove any irrelevant citations that are present.

To fine-tune an LLMs to develop the aforementioned abilities, we first need to construct training data. The most straightforward idea is to create a set of responses with poor citation quality, each paired with a corresponding response that has perfect citation quality. We attempt to use the answers from WebGLM-QA dataset as positive responses and generate negative responses by randomly adding or deleting citations. Unfortunately, this approach proved ineffective, primarily because the citation quality in dataset is not high enough. An evaluation of the dataset's answers revealed that the citation recall and citation precision are only 73.77% and 69.50%, respectively, which doesn't even reach the citation quality found in the responses generated by fine-tuned open-source models.

Due to this issue, we had to rely on an NLI model to help us construct high-quality target responses. We split the original responses from the dataset into statements. For each statement, after removing the existing citations, we enumerate all possible combinations of citations and use an NLI model to determine if each combination supports the statement. Then, we incorporate all the gold citations into the dataset. By following these operations, we obtain a dataset containing four fields: question, references, statement, and target citations, which will be used for training the refiner.

To avoid altering the original text of the answers, we only need the refiner to output the ids of the references that the statement should actually cite. Since the refiner outputs ids rather than complete statements, the additional computational overhead of applying this method to RAG scenario is actually minimal. Moreover, in our Generate-then-Refine approach, the generating and refining are decoupled, allowing the refiner to enhance citation quality in responses generated by any method.

### 4.2    Results

In this section, we fine-tune a Mistral-7B model [30] to serve as the refiner. The main results are summarized in Tables 4, 5, 6, and we have the key observations as follows.

First, whether through few-shot or fine-tuning, the responses generated by the model can achieve significant improvements in citation quality after refining. All models show improvements in citation F1 across three datasets. In the few-shot experiments on WebGLM-QA, Llama2 series models initially perform poorly but achieve improvements of 22.18% and 23.73% respectively with the

**Table 4.** Experiments on WebGLM-QA.

| Models | Citation (ALCE) | | | Citation (Ours) | | |
|---|---|---|---|---|---|---|
| | Recall | Precision | F1 | Recall | Precision | F1 |
| **Few-Shot + Refine** | | | | | | |
| gpt-3.5-turbo | **75.20**(+0.99) | 81.81(+6.90) | **78.37**(+3.81) | 80.40(+2.09) | **88.54**(+10.36) | **84.27**(+6.03) |
| llama2-7b | 49.02(+21.17) | 77.28(+23.04) | 59.99(+23.19) | 51.03(+22.08) | 74.76(+17.43) | 60.66(+22.18) |
| llama2-13b | 52.30(**+22.22**) | 77.89(**+26.95**) | 62.58(**+24.76**) | 55.09(**+23.31**) | 76.88(**+21.25**) | 64.19(**+23.73**) |
| mistral-7b | 73.33(+0.30) | **83.53**(+17.14) | 78.10(+8.55) | 75.81(+1.50) | 83.67(+20.83) | 79.55(+11.45) |
| llama3-8b | 69.28(-3.60) | 81.56(+8.57) | 74.92(+1.99) | 72.58(-2.00) | 86.01(+16.76) | 78.73(+6.91) |
| glm4-9b | 67.76(+6.78) | 83.31(+9.37) | 74.73(+7.90) | 70.26(+8.15) | 83.62(+11.53) | 76.36(+9.63) |
| qwen2-7b | 71.11(+8.04) | 82.11(+15.56) | 76.22(+11.45) | 72.98(+9.42) | 80.78(+15.26) | 76.68(+12.16) |
| **Fine-Tuning + Refine** | | | | | | |
| llama2-7b | 77.74(-1.22) | **89.03**(+9.19) | 83.00(**+3.61**) | 79.25(**+0.08**) | 89.79(+13.85) | 84.19(**+6.67**) |
| llama2-13b | 78.01(-1.55) | 88.89(+5.97) | 83.10(+1.89) | 79.72(-0.49) | 90.17(+11.96) | 84.62(+5.43) |
| mistral-7b | 75.99(-3.53) | 87.52(+6.00) | 81.35(+0.84) | 78.40(-1.73) | 89.04(+11.65) | 83.38(+4.65) |
| llama3-8b | 78.08(-1.92) | 88.96(+5.98) | 83.17(+1.70) | 79.36(-1.11) | **90.21**(+12.41) | 84.44(+5.33) |
| glm4-9b | **78.20**(-0.87) | 88.97(+5.92) | **83.24**(+2.23) | **80.01**(+0.01) | 90.02(+12.08) | **84.72**(+5.76) |
| qwen2-7b | 72.97(-4.54) | 87.62(+7.16) | 79.63(+0.67) | 75.01(-3.47) | 89.47(**+16.57**) | 81.60(+6.02) |

**Table 5.** Experiments on ASQA.

| Models | Citation (ALCE) | | | Citation (Ours) | | |
|---|---|---|---|---|---|---|
| | Recall | Precision | F1 | Recall | Precision | F1 |
| **Few-Shot + Refine** | | | | | | |
| gpt-3.5-turbo | 53.82(+1.26) | 55.77(+3.90) | 54.78(+2.56) | 57.42(+3.46) | **87.10**(+19.53) | 69.23(+9.21) |
| llama2-7b | 42.29(**+22.54**) | 67.33(**+33.86**) | 51.95(**+27.11**) | 46.12(+24.73) | 75.66(**+33.81**) | 57.31(**+29.00**) |
| llama2-13b | 43.94(+21.69) | 68.20(+31.29) | 53.45(+25.68) | 48.65(**+25.03**) | 74.57(+31.76) | 58.88(+28.44) |
| mistral-7b | 60.09(+3.04) | **72.85**(+15.08) | 65.86(+8.45) | 65.54(+5.40) | 82.44(+23.79) | 73.02(+13.64) |
| llama3-8b | 65.40(-1.98) | 68.09(+3.32) | 66.72(+0.67) | 71.26(+2.73) | 84.39(+16.92) | 77.27(+9.28) |
| glm4-9b | 64.63(+6.13) | 70.51(+8.67) | **67.44**(+7.32) | 69.15(+9.30) | 82.12(+16.76) | 75.08(+12.60) |
| qwen2-7b | **65.80**(+9.05) | 66.23(+8.63) | 66.01(+8.84) | **73.63**(+16.36) | 82.84(+19.88) | **77.96**(+17.98) |
| **Fine-Tuning + Refine** | | | | | | |
| llama2-7b | 65.14(+5.12) | 69.04(+22.32) | 67.03(+14.49) | 74.09(**+12.48**) | 84.82(+33.44) | 79.09(+23.06) |
| llama2-13b | 65.88(+5.53) | 69.32(+24.06) | 67.56(+15.83) | **74.32**(+11.39) | 84.58(+33.87) | 79.12(+22.96) |
| mistral-7b | 66.11(**+6.03**) | 73.38(+24.19) | 69.56(+15.46) | 72.51(+10.50) | **85.78**(+32.98) | 78.59(+21.55) |
| llama3-8b | **68.28**(+4.66) | **74.17**(+24.59) | **71.10**(+15.37) | 74.20(+10.18) | 85.58(+32.65) | **79.48**(+21.54) |
| glm4-9b | 65.89(+4.25) | 72.21(**+30.44**) | 68.91(**+19.11**) | 71.47(+9.44) | 84.41(**+40.39**) | 77.40(**+25.91**) |
| qwen2-7b | 60.56(+2.30) | 69.02(+25.02) | 64.51(+14.38) | 67.47(+6.91) | 83.96(+37.36) | 74.82(+22.15) |

help of the refiner, narrowing the performance gap with other models. On the other two datasets, Llama2 series models also achieved improvements of nearly 29% and 18%.

Second, refiner exhibits excellent generalization. Previous experiments found that fine-tuning method doesn't exhibit strong generalization, as the model's capabilities are constrained by data distribution. For instance, the fine-tuned

**Table 6.** Experiments on ELI5.

| Models | Citation (ALCE) | | | Citation (Ours) | | |
|---|---|---|---|---|---|---|
| | Recall | Precision | F1 | Recall | Precision | F1 |
| **Few-Shot + Refine** | | | | | | |
| gpt-3.5-turbo | 23.69(+0.36) | 39.54(+14.67) | 29.63(+5.55) | 26.11(+1.10) | 72.33(+25.09) | 38.37(+5.66) |
| llama2-7b | 28.34(+12.52) | 61.56(+22.43) | 38.81(+16.28) | 29.74(+12.98) | 70.27(+28.13) | 41.79(+17.81) |
| llama2-13b | 27.08(+11.68) | 61.35(+28.47) | 37.57(+16.60) | 28.77(+12.08) | 71.77(+34.11) | 41.07(+17.95) |
| mistral-7b | 45.27(+1.54) | 60.69(+19.84) | 51.86(+9.62) | 48.36(+3.33) | 74.75(+28.80) | 58.73(+13.24) |
| llama3-8b | 42.94(-0.04) | 55.40(+8.92) | 48.38(+3.72) | 47.54(+2.45) | 76.86(+23.98) | 58.74(+10.07) |
| glm4-9b | 35.47(+5.88) | 61.21(+16.67) | 44.91(+9.36) | 38.03(+7.08) | 72.64(+23.75) | 49.92(+12.02) |
| qwen2-7b | 42.79(+7.39) | 60.70(+19.00) | 50.20(+11.90) | 45.06(+9.10) | 70.22(+24.08) | 54.89(+14.48) |
| **Fine-Tuning + Refine** | | | | | | |
| llama2-7b | 48.32(-1.12) | 66.62(+14.96) | 56.01(+5.49) | 51.19(+1.26) | 80.33(+25.20) | 62.53(+10.13) |
| llama2-13b | 47.92(+0.91) | 66.19(+15.19) | 55.59(+6.67) | 51.28(+3.40) | 80.67(+24.21) | 62.70(+10.88) |
| mistral-7b | 48.23(-0.69) | 66.14(+11.68) | 55.78(+4.24) | 51.25(+1.43) | 80.38(+22.96) | 62.59(+9.24) |
| llama3-8b | 48.69(-0.02) | 65.24(+12.62) | 55.76(+5.17) | 51.65(+2.30) | 79.13(+21.98) | 62.50(+9.54) |
| glm4-9b | 48.82(+0.13) | 67.43(+14.73) | 56.64(+6.02) | 52.18(+2.34) | 79.82(+24.75) | 63.11(+10.78) |
| qwen2-7b | 43.45(-1.36) | 63.91(+11.29) | 51.73(+3.33) | 46.74(+0.80) | 79.36(+24.95) | 58.83(+9.01) |

LLMs don't perform as well on ASQA compared to few-shot methods. However, after refining, all six fine-tuned LLMs achieved over a 20% increase in citation F1 on ASQA. Except for Qwen2-7b, the other five fine-tuned models surpassed the results of few-shot. These results indicate that changes in the distribution of data have minimal impact on the refiner.

Third, refiner primarily enhances citation quality through improved citation precision. In many results, citation recall has decreased actually, but the increase in citation precision is substantial. For example, the fine-tuned glm4-9b model achieved a staggering 40% increase in citation precision on ASQA. This illustrates that the refiner effectively captures the relationship between statements and references, accurately determining whether a reference truly supports the response.

### 4.3   Additional Evaluation

To further demonstrate the effectiveness of our proposed Generate-then-Refine method, we proceed with additional evaluations. We need to confirm whether the improvement in citation quality is genuine or merely aligned with NLI model's preferences since we use an NLI model to determine the gold citation while constructing the training data for the refiner and also use NLI model in the evaluation.

Thus, we replace NLI model with GPT-3.5-turbo when measuring citation recall and citation precision. We request GPT to output 'Yes' only if it believes the cited references supports the statement, otherwise output 'No'. We conduct experiments with Llama2-7b and Llama2-8b models on the test set of WebGLM-QA.

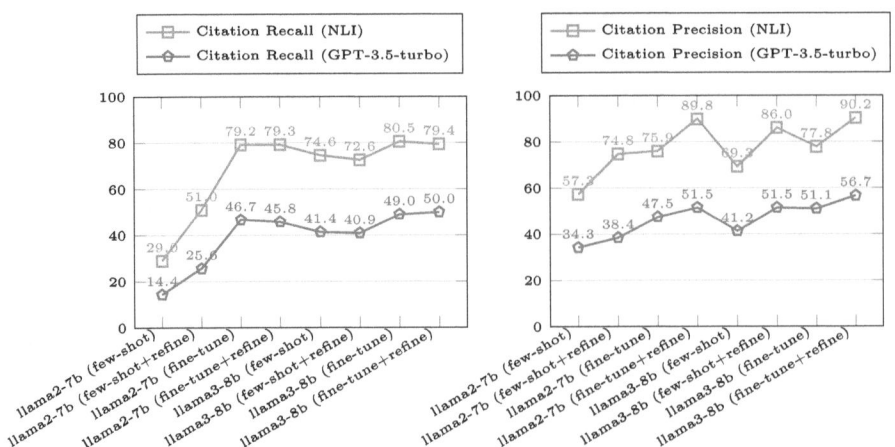

**Fig. 1.** Evaluation results using NLI model and GPT-3.5-turbo.

The experiment results as shown in Fig. 1. In the evaluation of citation recall and citation precision metrics, there is a significant difference in the judgment criteria between NLI model and GPT-3.5-turbo model. However, the evaluation results from both models show a noticeable positive correlation. From the perspective of evaluating the relative quality of citations, their evaluation results are consistent. In other words, responses with higher citation quality as measured by NLI model are also recognized by GPT-3.5-turbo model. This indicates that the improvement in citation quality brought about by our proposed method is not due to the NLI model's preference. The improvements in citation quality brought by the refiner are genuine.

Meanwhile, we also attempted to guide the LLM to become an excellent refiner using the few-shot method, which would simplify the pipeline if effective. Unfortunately, none of the LLMs we tried inherently possessed strong refining capabilities, and using the few-shot method for refining significantly reduced citation quality.

Our proposed Generate-then-Refine method combines both pre-hoc and post-hoc citation. To obtain more comprehensive experiment results, we removed the generated citations and re-added them using a rule-based post-hoc method. We still conduct experiments with the Llama2-7b and Llama3-8b models on the WebGLM-QA test set. We used BLEU-4 [27] and ROUGE-L [26] metrics to match the statements in the answers with those in the references, setting the threshold at 0.3. Whenever the similarity score exceeded the threshold, we added a citation to the corresponding answer statement.

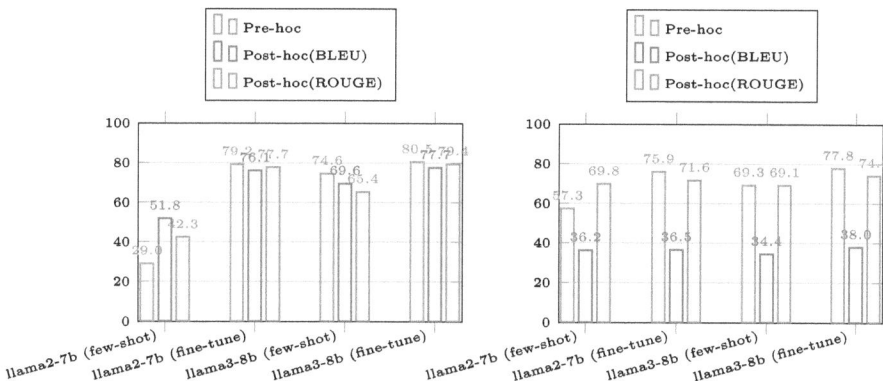

**Fig. 2.** Comparison of Pre-hoc Citation and Post-hoc Citation. On the left is the evaluation of Citation Recall, and on the right is the evaluation of Citation Precision.

The experiment results as shown in Fig. 2. From the experiment results, it can be observed that Post-hoc method only works for answers generated by models that lack attribution capabilities. Once a model has good attribution capabilities, the Post-hoc method performs worse than Pre-hoc method in both Citation Recall and Citation Precision metrics. Moreover, due to the difficulty in setting a similarity threshold, the BLEU-based method performs significantly worse than the ROUGE-based method in Citation Precision metric.

## 5    Conclusion

In this work, we comprehensively evaluate the ability of the latest LLMs to generate citations in their responses. We introduce new citation evaluation metrics to address shortcomings in the existing evaluation framework. To improve the citation quality in LLMs' responses, we propose Generate-then-Refine method, which fine-tunes a model to serve as a refiner. Our experiments show that our method substantially improves the quality of citations.

**Acknowledgements.** This work was funded by the National Natural Science Foundation of China (NSFC) under Grants No. 62372431 and 62472408, the Strategic Priority Research Program of the CAS under Grants No. XDB0680102, XDB0680301, the National Key Research and Development Program of China under Grants No. 2023YFA1011602, the Youth Innovation Promotion Association CAS under Grants No. 2021100, the Lenovo-CAS Joint Lab Youth Scientist Project, and the project under Grants No. JCKY2022130C039.

# References

1. Brown, T., Mann, B., Ryder, N., Subbiah, M., Kaplan, J., et al.: Language models are few-shot learners. In: Advances in Neural Information Processing Systems, pp. 1877–1901 (2020)
2. Ji, Z., Lee, N., Frieske, R., Yu, T., Su, D., et al.: Survey of hallucination in natural language generation. ACM Comput. Surv. **55**(12), 1–38 (2023)
3. Lewis, P., Perez, E., Piktus, A., Petroni, F., Karpukhin, V., et al.: Retrieval-augmented generation for knowledge-intensive nlp tasks. In: Advances in Neural Information Processing Systems, pp. 9459–9474 (2020)
4. Shuster, K., Poff, S., Chen, M., Kiela, D., Weston, J.: Retrieval augmentation reduces hallucination in conversation. arXiv preprint arXiv:2104.07567 (2021)
5. Gao, Y., Xiong, Y., Gao, X., Jia, K., Pan, J., et al.: Retrieval-augmented generation for large language models: a survey. arXiv preprint arXiv:2312.10997 (2023)
6. Jiang, Z., Xu, F. F., Gao, L., Sun, Z., Liu, Q., et al.: Active retrieval augmented generation. arXiv preprint arXiv:2305.06983 (2023)
7. Shao, Z., Gong, Y., Shen, Y., Huang, M., Duan, N., et al.: Enhancing retrieval-augmented large language models with iterative retrieval-generation synergy. arXiv preprint arXiv:2305.15294 (2023)
8. Gao, T., Yen, H., Yu, J., Chen, D.: Enabling large language models to generate text with citations. arXiv preprint arXiv:2305.14627 (2023)
9. Huang, J., Chang, K. C. C.: Citation: A Key to Building Responsible and Accountable Large Language Models. arXiv preprint arXiv:2307.02185 (2023)
10. Nakano, R., Hilton, J., Balaji, S., Wu, J., Ouyang, L., et al.: Webgpt: browser-assisted question-answering with human feedback. arXiv preprint arXiv:2112.09332 (2021)
11. Liu, X., Lai, H., Yu, H., Xu, Y., Zeng, A., et al.: WebGLM: Towards an efficient web-enhanced question answering system with human preferences. In: Proceedings of the 29th ACM SIGKDD Conference on Knowledge Discovery and Data Mining, pp. 4549–4560 (2023)
12. Stelmakh, I., Luan, Y., Dhingra, B., Chang, M. W.: ASQA: factoid questions meet long-form answers. arXiv preprint arXiv:2204.06092 (2022)
13. Fan, A., Jernite, Y., Perez, E., Grangier, D., Weston, J., et al.: ELI5: Long form question answering. arXiv preprint arXiv:1907.09190 (2019)
14. Menick, J., Trebacz, M., Mikulik, V., Aslanides, J., Song, F., et al.: Teaching language models to support answers with verified quotes. arXiv preprint arXiv:2203.11147 (2022)
15. Qian, H., Zhu, Y., Dou, Z., Gu, H., Zhang, X., et al.: Webbrain: learning to generate factually correct articles for queries by grounding on large web corpus. arXiv preprint arXiv:2304.04358 (2023)
16. Qin, Y., Cai, Z., Jin, D., Yan, L., Liang, S., et al.: Webcpm: interactive web search for Chinese long-form question answering. arXiv preprint arXiv:2305.06849 (2023)
17. Sun, H., Cai, H., Wang, B., Hou, Y., Wei, X., et al.: Towards verifiable text generation with evolving memory and self-reflection. arXiv preprint arXiv:2312.09075 (2023)
18. Huang, C., Wu, Z., Hu, Y., Wang, W.: Training language models to generate text with citations via fine-grained rewards. arXiv preprint arXiv:2402.04315 (2024)
19. Rashkin, H., Nikolaev, V., Lamm, M., Aroyo, L., Collins, M., et al.: Measuring attribution in natural language generation models. Comput. Linguist. **49**(4), 777–840 (2023)

20. Gao, L., Dai, Z., Pasupat, P., Chen, A., Chaganty, A. T., et al.: Rarr: researching and revising what language models say, using language models. arXiv preprint arXiv:2210.08726 (2022)
21. Bohnet, B., Tran, V. Q., Verga, P., Aharoni, R., Andor, D., et al.: Attributed question answering: Evaluation and modeling for attributed large language models. arXiv preprint arXiv:2212.08037 (2022)
22. Liu, N. F., Zhang, T., Liang, P.: Evaluating verifiability in generative search engines. arXiv preprint arXiv:2304.09848 (2023)
23. Yue, X., Wang, B., Chen, Z., Zhang, K., Su, Y., et al.: Automatic evaluation of attribution by large language models. arXiv preprint arXiv:2305.06311 (2023)
24. Kamalloo, E., Jafari, A., Zhang, X., Thakur, N., Lin, J.: Hagrid: A human-llm collaborative dataset for generative information-seeking with attribution. arXiv preprint arXiv:2307.16883 (2023)
25. Hu, N., Chen, J., Wu, Y., Qi, G., Bi, S., et al.: Benchmarking large language models in complex question answering attribution using knowledge graphs. arXiv preprint arXiv:2401.14640 (2024)
26. Lin, C. Y.: Rouge: a package for automatic evaluation of summaries. In: Text Summarization Branches Out, pp. 74–81 (2004)
27. Papineni, K., Roukos, S., Ward, T., Zhu, W. J.: Bleu: a method for automatic evaluation of machine translation. In: Proceedings of the 40th Annual Meeting of the Association for Computational Linguistics, pp. 311–318 (2002)
28. OpenAI: Introducing ChatGPT. https://openai.com/blog/chatgpt. Accessed 13 June 2024
29. Touvron, H., Martin, L., Stone, K., Albert, P., Almahairi, A., et al.: Llama 2: open foundation and fine-tuned chat models. arXiv preprint arXiv:2307.09288 (2023)
30. Jiang, A. Q., Sablayrolles, A., Mensch, A., Bamford, C., Chaplot, D.S., et al.: Mistral 7B. arXiv preprint arXiv:2310.06825 (2023)
31. Meta: Introducing Meta Llama 3: The most capable openly available LLM to date. https://ai.meta.com/blog/meta-llama-3. Accessed 13 June 2024
32. GLM Team, et al.: ChatGLM: A Family of Large Language Models from GLM-130B to GLM-4 All Tools. arXiv preprint arXiv:2406.12793 (2024)
33. Yang, A., Yang, B., Hui, B., Zheng, B., Yu, B., et al.: Qwen2 Technical Report. arXiv preprint arXiv:2407.10671 (2024)
34. Honovich, O., Aharoni, R., Herzig, J., Taitelbaum, H., Kukliansy, D., et al.: TRUE: re-evaluating factual consistency evaluation. arXiv preprint arXiv:2204.04991 (2022)
35. Hu, E. J., Shen, Y., Wallis, P., Allen-Zhu, Z., Li, Y., et al.: Lora: low-rank adaptation of large language models. arXiv preprint arXiv:2106.09685 (2021)

# Are Large Language Models More Honest in Their Probabilistic or Verbalized Confidence?

Shiyu Ni[1,2], Keping Bi[1,2], Lulu Yu[1,2], and Jiafeng Guo[1,2(✉)]

[1] CAS Key Lab of Network Data Science and Technology, ICT, CAS, Beijing, China
{nishiyu23z,bikeping,yululu23s,guojiafeng}@ict.ac.cn
[2] University of Chinese Academy of Sciences, Beijing, China

**Abstract.** Large language models (LLMs) have been found to produce hallucinations when the question exceeds their internal knowledge boundaries. A reliable model should have a clear perception of its knowledge boundaries, providing correct answers within its scope and refusing to answer when it lacks knowledge. Existing research on LLMs' perception of their knowledge boundaries typically uses either the probability of the generated tokens or the verbalized confidence as the model's confidence in its response. However, these studies overlook the differences and connections between the two. In this paper, we conduct a comprehensive analysis and comparison of LLMs' probabilistic perception and verbalized perception of their factual knowledge boundaries. First, we investigate the pros and cons of these two perceptions. Then, we study how they change under questions of varying frequencies. Finally, we measure the correlation between LLMs' probabilistic confidence and verbalized confidence. Experimental results show that 1. LLMs' probabilistic perception is generally more accurate than verbalized perception but requires an in-domain validation set to adjust the confidence threshold. 2. Both perceptions perform better on less frequent questions. 3. It is challenging for LLMs to accurately express their internal confidence in natural language.

**Keywords:** LLMs' perception of their knowledge boundaries ·
Probabilistic perception · Verbalized perception

## 1 Introduction

Recently, large language models (LLMs) have demonstrated remarkable performance across various NLP tasks [3,6,18]. Despite their impressive capabilities, LLMs have several significant limitations. One critical issue is that LLMs can produce hallucinations, generating factually incorrect answers that appear accurate, primarily occurring when the question exceeds the model's internal knowledge boundaries [23]. A reliable system should provide correct answers when it knows the answer and refuses to answer when it does not, rather than fabricating unreliable responses, which is especially important in areas such as safety and healthcare. This requires the model to have a clear understanding of its knowledge boundaries, knowing what it knows and what it does not know.

A model with a clear perception of its knowledge boundaries is not only more reliable but can also aid downstream tasks. For example, it can help retrieval augmentation (RA) where RA can be triggered only when the model expresses uncertainty about its answers to enhance efficiency and effectiveness, which we call adaptive retrieval augmentation. This is because retrieval augmentation incurs additional overhead and the quality of retrieved documents cannot be guaranteed, potentially misleading the model instead.

Existing research on the model's perception of its knowledge boundaries mainly involves two types of confidence: probabilistic confidence [4,7,9,10,19] where they use the probability of the generated tokens as the model's confidence and verbalized confidence where LLMs are taught to express their confidence in words [14,17,20–22,24]. These represent the model's probabilistic and verbalized perceptions of its knowledge boundaries. However, these works only explore these perspectives separately, overlooking their differences and connections.

In this paper, we investigate LLMs' probabilistic perception and verbalized perception of their factual knowledge boundaries, analyzing the differences and correlations between them. Specifically, we try to answer three research questions. **RQ1: What are the pros and cons of these two perceptions?** Inspired by the previous finding that LLMs can generate more accurate answers for more common questions [15], we also wonder **RQ2: How do these two perceptions change under questions of varying frequencies?** In addition to exploring the differences between these two perceptions, we also study their correlations, so the last research question is **RQ3: Can LLMs accurately express their internal confidence in natural language?**.

To answer **RQ1**, we choose four widely used LLMs and conduct experiments on the representative factual QA benchmark, i.e., Natural Questions (NQ) [12]. Experimental results indicate that LLMs' probabilistic perception of their knowledge boundaries is more accurate than their verbalized perception. However, probabilistic perception necessitates the use of an in-domain dataset to determine an appropriate confidence threshold for binarizing continuous probabilistic confidence. In contrast, LLMs' verbalized perception performs at a reasonable level without requiring additional setup.

To answer **RQ2**, we test two powerful black-box models on the Parent and Child dataset [2] where questions in the Child dataset are less common than those in the Parent dataset. We find that both LLMs' probabilistic perception and verbalized perception of their knowledge boundaries perform better on the Child dataset than on the Parent dataset. This indicates that LLMs' perception levels decline on more familiar questions. Additionally, for less common questions, probabilistic perception outperforms verbalized perception by a greater margin.

To answer **RQ3**, we adopt two commonly used correlation coefficient methods: the Spearman correlation coefficient [8] and the Kendall correlation coefficient [1]. These methods are used to calculate the correlation between probabilistic confidence and verbalized confidence for four LLMs (the same models used in RQ1) on the NQ, Parent, and Child datasets. We show that, overall, LLMs' verbalized confidence is positively correlated with their probabilistic confidence. However, at a finer granularity, the correlation is weak and varies significantly across different datasets. Therefore, we conclude that it is challenging for LLMs to accurately express their internal confidence in natural language.

## 2    Related Work

Many studies have investigated deep neural models' perception of their knowledge boundaries, which can be primarily divided into two categories: probabilistic perception and verbalized perception.

***Probabilistic Perception.*** This series of works [4,7,9,10,19] utilizes the probability of the generated tokens as the model's confidence in the answer, which we refer to as probabilistic confidence. It has been observed that deep neural networks tend to be overconfident, a problem that can be effectively mitigated by adjusting the generation temperature [7]. Subsequent research has examined the perception levels of pre-trained transformers. It has been found that BERT-style models are generally well-calibrated [4], whereas generative language models lack this level of perception [9]. More recent studies have explored LLMs' perception of their knowledge boundaries. Kadavath et al. [10] and Si et al. [19] demonstrate that using appropriate prompting techniques can make LLMs more reliable.

***Verbalized Perception.*** With the development of LLMs, several studies have demonstrated that LLMs can express their confidence in words [14,17,20–22,24], which we refer to as verbalized confidence. Lin et al. [14] first show that a model (GPT-3) can learn to express confidence about its answers in natural language. Then, Yin et al. [24] evaluate LLMs' self-knowledge by assessing whether they can identify unanswerable or unknowable questions. To enhance LLMs' ability to verbalize their confidence, some studies focus on prompting methods [20,21] while Yang et al. [22] are dedicated to training methods.

These studies only explore LLMs' perception of their knowledge boundaries from either a probabilistic or verbalized perspective, overlooking the differences and connections between them, which we investigate in this paper.

## 3    Preliminaries

In this section, we will introduce our task and the basic experimental setup.

### 3.1    Task Formulation

***Open-Domain QA***. The goal of open-domain QA is to ask the model $\mathcal{M}$ to provide an answer $a$ for a given question $q$. Unlike previous small-scale models [5] that rely on the retrieve-then-read pipeline [11,13,16], where relevant external documents are first retrieved for the question $q$ and then the model extracts the correct answer from these documents, LLMs can answer the question directly based on their internal knowledge. We instruct LLMs to answer using prompt $p$ and the format can be described as:

$$a = f_{\text{LLM}}(q, p) \tag{1}$$

***Confidence Elicitation***. Instead of only obtaining the answer, we also expect the model to express its confidence $c$ in the answer. We focus on two types of confidence: verbalized confidence and probabilistic confidence.

**Verbalized Confidence.** LLMs are found to have the power to express their confidence in words which we refer to as verbalized confidence $c_{verb}$. We use $\hat{p}$ to instruct the model to generate the answer along with its verbalized confidence and the format is:

$$(a, c_{verb}) = f_{\text{LLM}}(q, \hat{p}) \tag{2}$$

where $c_{verb} = 1$ indicates the model is confident in its answer while $c_{verb} = 0$ means the opposite.

**Probabilistic Confidence.** Perplexity can reflect the model's internal degree of certainty in the answer which we refer to as probabilistic confidence $c_{prob}$. For an answer $a$ consisting of $n$ tokens $\{a_1, a_2, \cdots a_n\}$, $c_{prob}$ is computed as:

$$c_{prob} = -\frac{1}{n} \sum_{i=1}^{n} \log P(a_i | a_{<i}) \tag{3}$$

where a lower $c_{prob}$ implies that the model is more confident in the answer.

### 3.2  Experimental Setup

***Prompts.*** To facilitate the model answering the question and expressing its confidence in the answer, we use prompt $\hat{p} =$ "*Answer the following question based on your internal knowledge with one or few words. If you are sure the answer is accurate and correct, please say certain after the answer. If you are not confident with the answer, please say uncertain. Question: {question}. Answer:*" where {question} is the placeholder for the question $q$.

***Models.*** We conduct experiments on two representative open-source models (Llama2-7B-Chat and Mistral-7B-Instruct-v0.2) and two widely used black-box models that can return the probability of the generated tokens, including Chat-GPT (gpt-3.5-turbo-1106) and GPT-Instruct (gpt-3.5-turbo-instruct). For all the models, we set temperature $= 1$ to obtain the raw probabilities.

## 4  Evaluating LLMs' Probabilistic and Verbalized Perceptions of Their Knowledge Boundaries

In this section, we investigate the performance of LLMs' probabilistic and verbalized perceptions of their knowledge boundaries and try to answer **RQ1**.

### 4.1   Exprimental Setup

***Datasets.*** We conduct experiments on a widely used open-domain QA dataset, Natural Questions (NQ) [12]. NQ is constructed using Google Search queries with annotated short or long answers related to factual knowledge. For our experiments, we use only questions with short answers from the test set and treat these short answers as labels. We randomly sample 20% of the data as the validation set, and report results on the remaining data.

***Metrics.*** Following previous research [17], we use **Alignment, Overconfidence**, and **Conservativeness** to measure LLMs' perception of their knowledge boundaries. **Accuracy** (acc for short) is employed to represent the QA performance, where a response is deemed correct if it contains the ground-truth label. **Uncertain rate** is computed as the proportion of samples where the model expresses uncertainty and is used to represent the model's uncertainty level.

In view of verbalized perception, $Alignment_{verb}$ is computed by the proportion of samples where LLMs' confidence matches the correctness of the response (i.e., $c_{verb} = acc$). $Overconfidence_{verb}$ is the proportion of samples where the model is confident but the response is incorrect (i.e., $c_{verb} = 1, acc = 0$), and $Conservativenes_{verb}$ is used to measure the proportion of samples where the model expresses uncertainty but the response is correct (i.e., $c_{verb} = 0, acc = 1$).

Unlike verbalized confidence, probabilistic confidence is a continuous value and cannot be directly matched with binary accuracy. Therefore, we set a threshold $\lambda$ to binarize probabilistic confidence and the format is:

$$c_{prob} = \begin{cases} 1, & \text{if } c_{prob} \leq \lambda \\ 0, & \text{if } c_{prob} > \lambda \end{cases} \tag{4}$$

Then, similar to the metrics for verbalized perception, we compute $Alignment_{prob}$, $Overconfidence_{prob}$, and $Conservativeness_{prob}$. We use the confidence threshold which achieves the optimal $Alignment_{prob}$ on the validation set as $\lambda$.

### 4.2   Results and Analysis

The results of LLMs' QA performance and probabilistic and verbalized perceptions of their knowledge boundaries are shown in Table 1. We observe that: 1) When expressing confidence in words, LLMs are not well-calibrated and tend to be overconfident which is consistent with the previous findings [17]. 2) The probabilistic confidence is consistently much lower than the verbalized confidence, and the probabilistic alignment is significantly higher than the verbalized alignment across all models. This indicates that, compared to judging the correctness of an answer in words, LLMs have a better probabilistic perception of their knowledge boundaries. The possible reason may be that, when expressing confidence in words, LLMs do not have access to the probability distribution of the generated answer, which can be a useful signal representing the correctness of the answer. 3) A good probabilistic perception requires an additional

**Table 1.** LLMs' probabilistic and verbalized perceptions of their knowledge boundaries of LLMs on NQ. Bold denotes the highest score for each model. Unc., Conserv., and Overconf. stand for Uncertain rate, Conservativeness, and Overconfidence, respectively.

| Model | Acc | Strategy | Unc. | Alignment | Overconf. | Conserv. |
|---|---|---|---|---|---|---|
| Llama2 | 0.2957 | Verb. | 0.1894 | 0.4512 | **0.5319** | 0.0170 |
| | | Prob. | **0.8764** | **0.7254** | 0.0512 | **0.2233** |
| Mistral | 0.2985 | Verb. | 0.4848 | 0.6260 | **0.2954** | 0.0786 |
| | | Prob. | **0.9034** | **0.7185** | 0.0398 | **0.2417** |
| GPT-Instruct | 0.4021 | Verb. | 0.1868 | 0.5182 | **0.4464** | 0.0354 |
| | | Prob. | **0.6891** | **0.6551** | 0.1269 | **0.2180** |
| ChatGPT | 0.4229 | Verb. | 0.2111 | 0.5252 | **0.4204** | 0.0554 |
| | | Prob. | **0.6443** | **0.6741** | 0.1294 | **0.1985** |

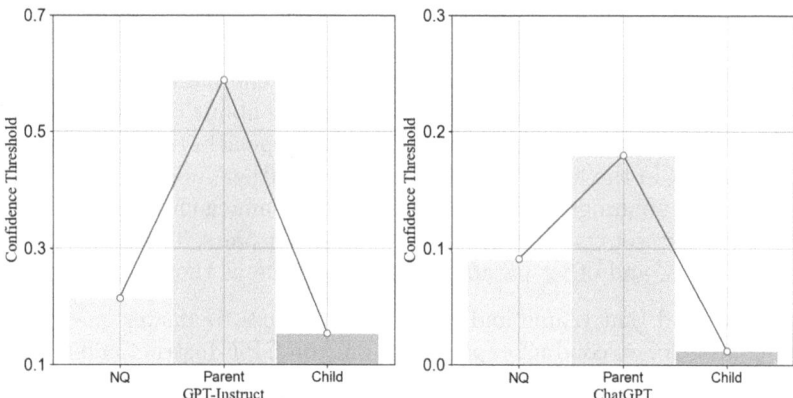

**Fig. 1.** The best threshold $\lambda$ for GPT-Instruct and ChatGPT on each dataset.

in-domain dataset to select an appropriate threshold. Figure 1 shows that the best probabilistic confidence of each model varies significantly across different datasets.

## 5   Effects of Question Frequency

LLMs often achieve better QA performance on common questions compared to unfamiliar ones [15]. In this section, we investigate the effects of question frequency on LLMs' perception of their knowledge boundaries and answer **RQ2**.

### 5.1   Experimental Setup

***Datasets.*** The Parent-Child dataset is a collection of facts about actual celebrities and their parents, presented in the form "A's parent is B" and "B's child is

**Table 2.** LLMs' probabilistic and verbalized perceptions of their knowledge boundaries on Parent and Child datasets. Bold denotes the highest score for each model. Unc., Conserv., and Overconf. stand for Uncertain rate, Conservativeness, and Overconfidence respectively.

| Model | Dataset | Acc | Strategy | Unc. | Alignment | Overconf. | Conserv. |
|-------|---------|-----|----------|------|-----------|-----------|----------|
| GPT-Instruct | Parent | **0.5475** | Verb. | 0.3642 | 0.6424 | 0.2230 | **0.1346** |
| | | | Prob. | 0.2114 | 0.6846 | **0.2783** | 0.0372 |
| | Child | 0.1540 | Verb. | 0.6018 | 0.7243 | 0.2599 | 0.0157 |
| | | | Prob. | **0.8609** | **0.8593** | 0.0629 | 0.0778 |
| ChatGPT | Parent | **0.5778** | Verb. | 0.2288 | 0.6198 | 0.2868 | 0.0934 |
| | | | Prob. | 0.3877 | 0.7670 | 0.1337 | **0.0992** |
| | Child | 0.1322 | Verb. | 0.5632 | 0.6803 | **0.3121** | 0.0075 |
| | | | Prob. | **0.9339** | **0.8820** | 0.0259 | 0.0921 |

A" where A is the name of the celebrity and B is the name of A's parent [2]. LLMs are often more familiar with questions asking about the names of famous people's parents because these are more likely to appear in training corpora. We collect questions in the form "Who is A's mother/father" and name this dataset "Parent." Correspondingly, questions in the form "Name a child of B" are named "Child." Each dataset contains 1513 question-answer pairs. The dataset splitting strategy, metrics, and other parameters are the same as those in Sect. 4.

***Models.*** We find that Llama and Mistral often refuse to answer questions containing names. So, we conduct experiments only on GPT-Instruct and ChatGPT.

### 5.2   Results and Analysis

Table 2 shows the QA performance of GPT-Instruct and ChatGPT on questions of different frequencies, along with the models' perception of their knowledge boundaries. We find that: 1) LLMs achieve better QA performance which aligns with the previous findings in [2] and are more confident in the parent dataset compared to the child dataset. 2) Both verbalized and probabilistic alignment are higher on the child dataset. This indicates that LLMs have a better perception of their knowledge boundaries on less common questions rather than more familiar ones. 3) From common questions to unfamiliar ones, probabilistic alignment demonstrates a greater increase compared to verbalized alignment. The reason is that probabilistic confidence drops to a very low level on the child dataset which mitigates the level of overconfidence. It shows that these models have a very clear probabilistic understanding of what they do not know on the unfamiliar questions. At the same time, they also maintain the perception level of what they know. In view of verbalized perception, LLMs have a more accurate judgment of what they know. However, they are still overconfident which is the primary reason for the unsatisfactory perception of knowledge boundaries.

# 6   The Correlation Between LLMs' Probabilistic Confidence and Verbalized Confidence

In this section, we study the correlation between LLMs' probabilistic confidence and verbalized confidence to answer **RQ3**.

## 6.1   Experimental Setup

For a probabilistic confidence list $\mathcal{C}_p = \{c_p^1, c_p^2, \cdots c_p^n\}$ and a verbalized confidence list $\mathcal{C}_v = \{c_v^1, c_v^2, \cdots c_v^n\}$, we utilize two commonly used correlation coefficients: Spearman [8] and Kendall [1] correlation coefficients to measure their correlations.

***Spearman Coefficient.*** Spearman's rank correlation coefficient uses the rank of each value in the lists $\mathcal{C}_p$ and $\mathcal{C}_v$ to measure their correlation and the formula is:

$$\rho_s = 1 - \frac{6 \sum d_i^2}{n\left(n^2 - 1\right)} \tag{5}$$

where $d_i = rank_{c_p^i} - rank_{c_v^i}$ and $n$ is data count. The value of $\rho_s$ ranges from $[-1, 1]$, where a larger absolute value indicates a stronger correlation. 1 represents a perfect positive correlation, while $-1$ represents a perfect negative correlation.

***Kendall Coefficient.*** Kendall's rank correlation coefficient is defined based on the concepts of concordant pairs and discordant pairs. A concordant pair or discordant pair refers to a pair where the relative ordering of the two variables is consistent (e.g., $c_p^i > c_p^j$ and $c_v^i > c_v^j$) or not (e.g., $c_p^i > c_p^j$ and $c_v^i < c_v^j$). The format is:

$$\rho_k = \frac{e - f}{e + f} = \frac{e - f}{\frac{1}{2} \cdot n \cdot (n - 1)} \tag{6}$$

where $e$ is the count of concordant pairs, $f$ is the count of discordant pairs, and $n$ is the data count. Similar to $\rho_s$, the value of $\rho_k$ also ranges from $[-1, 1]$, where 1 represents a perfect positive correlation, while $-1$ represents the verse vice.

***Mode.*** We calculate the correlation coefficients in two modes. 1) **Vanilla**: We take LLMs' probabilistic confidence and verbalized confidence for all the data as $\mathcal{C}_p$ and $\mathcal{C}_v$. 2) **Bin-k**. To mitigate the influence of the order of individual samples and estimate the overall trend, we sort all the data in ascending order based on the probabilistic confidence and divide them into $k$ bins with the same length. The probabilistic confidence and verbalized confidence of each bin are the average values of the data within that bin. This yields lists $\mathcal{C}_p$ and $\mathcal{C}_v$, each of length $k$. In this paper, We set $k$ to 10.

## 6.2   Results and Analysis

The results can be seen in Table 3 and we visualize the changes in verbalized confidence with probabilistic confidence under the Bin-10 mode, as shown in Fig. 2. We conclude that: 1) In Vanilla mode, the correlation coefficient is small, but it is

**Table 3.** Correlation coefficients between LLMs' verbalized confidence and probabilistic confidence. Bold denotes the highest score on each dataset.

|  |  | Llama | Mistral | GPT-Instruct | | | ChatGPT | | |
|---|---|---|---|---|---|---|---|---|---|
|  |  | NQ | NQ | NQ | Parent | Child | NQ | Parent | Child |
| Vanilla | Spearman | 0.24 | **0.37** | 0.23 | 0.13 | **0.38** | 0.22 | **0.2** | 0.28 |
|  | Kendall | 0.2 | **0.3** | 0.19 | 0.1 | **0.31** | 0.18 | **0.16** | 0.23 |
| Bin-10 | Spearman | 0.75 | 0.81 | 0.9 | 0.12 | **0.92** | **0.95** | **0.88** | 0.48 |
|  | Kendall | 0.73 | 0.73 | 0.78 | 0 | **0.85** | **0.87** | **0.78** | 0.33 |

much higher in Bin-10 mode (except for ChatGPT on the Parent dataset). This indicates that the correlation between the model's probabilistic confidence and verbalized confidence is relatively low although there is an overall trend showing that verbalized uncertainty increases as probabilistic uncertainty increases. 2) The correlation varies significantly across different datasets for the same model. For instance, probabilistic confidence and verbalized confidence of GPT-instruct show a clear overall trend on the child dataset, whereas, on the parent dataset, they are almost entirely unrelated. Therefore, it is challenging for LLMs to accurately express their internal confidence in words.

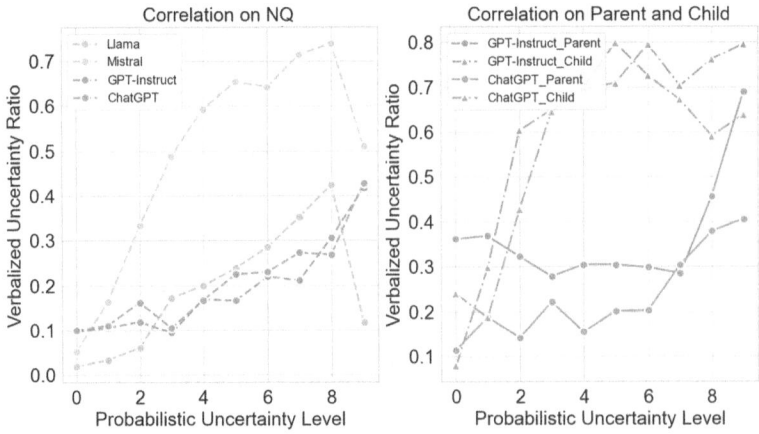

**Fig. 2.** Correlation between LLMs' probabilistic confidence and verbalized confidence. A higher uncertainty level means the model is less confident in its answer.

## Correlation Between LLMs' Confidence and Their QA Performance.

To more intuitively describe LLMs' perception of their knowledge boundaries, we investigate the correlation between probabilistic confidence and QA performance, as well as between verbalized confidence and QA performance, without binarizing the probabilistic confidence using a threshold. We calculate these correlations using Spearman and Kendall coefficients in both Vanilla and Bin-10 modes. The results are shown in Table 4. We observe that: 1) Both LLMs' probabilistic confidence and verbalized confidence are positively correlated with their QA performance, with probabilistic confidence generally showing a stronger correlation with QA performance than verbalized confidence. This indicates that LLMs have a better probabilistic perception of their knowledge boundaries even without a precise threshold, which aligns with the conclusion in Sect. 4. 2) Both correlations are very strong in the Bin-10 but relatively weak in the vanilla mode, representing that LLMs have a less nuanced but overall clear perception of their knowledge boundaries.

**Table 4.** Correlation coefficients between LLMs' confidence and their QA performance. Bold denotes the highest score for each correlation coefficient. Prob. and Verb. stand for probabilistic confidence and verbalized confidence, respectively.

| | | | Llama | Mistral | GPT-Instruct | | | ChatGPT | | |
|---|---|---|---|---|---|---|---|---|---|---|
| | | | NQ | NQ | NQ | Parent | Child | NQ | Parent | Child |
| Vanilla | Spearman | Prob. | **0.31** | 0.23 | **0.34** | **0.44** | 0.41 | **0.35** | **0.58** | **0.42** |
| | | Verb. | 0.22 | **0.28** | 0.22 | 0.26 | **0.44** | 0.19 | 0.18 | 0.37 |
| | Kendall | Prob. | **0.25** | 0.19 | **0.28** | **0.36** | 0.33 | **0.29** | **0.47** | 0.35 |
| | | Verb. | 0.22 | **0.28** | 0.22 | 0.26 | **0.44** | 0.19 | 0.18 | **0.37** |
| Bin-10 | Spearman | Prob. | **0.99** | **0.99** | 1 | **0.99** | **0.98** | 1 | **0.98** | 0.89 |
| | | Verb. | 0.7 | 0.91 | 0.7 | 0.88 | 0.85 | 0.45 | 0.81 | **0.92** |
| | Kendall | Prob. | **0.96** | **0.96** | 1 | **0.96** | **0.94** | 1 | **0.94** | 0.81 |
| | | Verb. | 0.62 | 0.81 | 0.62 | 0.78 | 0.73 | 0.35 | 0.71 | **0.85** |

## 7   Conclusion

In this paper, we conduct a comprehensive analysis and comparison of LLMs' probabilistic and verbalized perceptions of their factual knowledge boundaries. Specifically, we focus on answering three research questions: RQ1: What are the pros and cons of these two perceptions; RQ2: How do these two perceptions change under questions of varying frequencies; and RQ3: Can LLMs accurately express their internal confidence in natural language. We conduct extensive experiments on four commonly used LLMs and three open-source datasets and find that 1) LLMs' probabilistic perception is generally more accurate than verbalized perception but requires an in-domain validation set to adjust the confidence threshold. 2) Both perceptions perform better on less frequent questions and probabilistic perception outperforms verbalized perception by a greater margin on these questions. 3) It is challenging for LLMs to accurately express their internal confidence in natural language.

**Acknowledgements.** This work was funded by the National Natural Science Foundation of China (NSFC) under Grants No. 62302486, the Innovation Project of ICT CAS under Grants No. E361140, the CAS Special Research Assistant Funding Project, the Lenovo-CAS Joint Lab Youth Scientist Project, the project under Grants No. JCKY2022130C039, and the Strategic Priority Research Program of the CAS under Grants No. XDB0680102.

# References

1. Abdi, H.: The kendall rank correlation coefficient. Encyclopedia Measurement Stat. **2**, 508–510 (2007)
2. Berglund, L., et al.: The reversal curse: Llms trained on "a is b" fail to learn "b is a". arXiv preprint arXiv:2309.12288 (2023)
3. Brown, T., et al.: Language models are few-shot learners. Adv. Neural. Inf. Process. Syst. **33**, 1877–1901 (2020)
4. Desai, S., Durrett, G.: Calibration of pre-trained transformers. arXiv preprint arXiv:2003.07892 (2020)
5. Devlin, J., Chang, M.W., Lee, K., Toutanova, K.: Bert: pre-training of deep bidirectional transformers for language understanding. arXiv preprint arXiv:1810.04805 (2018)
6. Fan, R.Z., et al.: Reformatted alignment. arXiv preprint arXiv:2402.12219 (2024)
7. Guo, C., Pleiss, G., Sun, Y., Weinberger, K.Q.: On calibration of modern neural networks. In: International Conference on Machine Learning, pp. 1321–1330. PMLR (2017)
8. Hauke, J., Kossowski, T.: Comparison of values of pearson's and spearman's correlation coefficients on the same sets of data. Quaestiones Geographicae **30**(2), 87–93 (2011)
9. Jiang, Z., Araki, J., Ding, H., Neubig, G.: How can we know when language models know? on the calibration of language models for question answering. Trans. Assoc. Comput. Linguist. **9**, 962–977 (2021)
10. Kadavath, S., et al.: Language models (mostly) know what they know. arXiv preprint arXiv:2207.05221 (2022)
11. Karpukhin, V., et al.: Dense passage retrieval for open-domain question answering. arXiv preprint arXiv:2004.04906 (2020)
12. Kwiatkowski, T., et al.: Natural questions: a benchmark for question answering research. Trans. Assoc. Comput. Linguist. **7**, 453–466 (2019)
13. Lewis, P., et al.: Retrieval-augmented generation for knowledge-intensive nlp tasks. Adv. Neural. Inf. Process. Syst. **33**, 9459–9474 (2020)
14. Lin, S., Hilton, J., Evans, O.: Teaching models to express their uncertainty in words. arXiv preprint arXiv:2205.14334 (2022)
15. Mallen, A., Asai, A., Zhong, V., Das, R., Khashabi, D., Hajishirzi, H.: When not to trust language models: Investigating effectiveness of parametric and nonparametric memories. arXiv preprint arXiv:2212.10511 (2022)
16. Ni, S., Bi, K., Guo, J., Cheng, X.: A comparative study of training objectives for clarification facet generation. In: Proceedings of the Annual International ACM SIGIR Conference on Research and Development in Information Retrieval in the Asia Pacific Region, pp. 1–10 (2023)
17. Ni, S., Bi, K., Guo, J., Cheng, X.: When do llms need retrieval augmentation? mitigating llms' overconfidence helps retrieval augmentation. arXiv preprint arXiv:2402.11457 (2024)

18. Ouyang, L., et al.: Training language models to follow instructions with human feedback. Adv. Neural. Inf. Process. Syst. **35**, 27730–27744 (2022)
19. Si, C., et al.: Prompting gpt-3 to be reliable. arXiv preprint arXiv:2210.09150 (2022)
20. Tian, K., et al.: Just ask for calibration: strategies for eliciting calibrated confidence scores from language models fine-tuned with human feedback. arXiv preprint arXiv:2305.14975 (2023)
21. Xiong, M., et al.: Can llms express their uncertainty? An empirical evaluation of confidence elicitation in llms. arXiv preprint arXiv:2306.13063 (2023)
22. Yang, Y., Chern, E., Qiu, X., Neubig, G., Liu, P.: Alignment for honesty. arXiv preprint arXiv:2312.07000 (2023)
23. Yin, X., Zhang, X., Ruan, J., Wan, X.: Benchmarking knowledge boundary for large language model: a different perspective on model evaluation. arXiv preprint arXiv:2402.11493 (2024)
24. Yin, Z., Sun, Q., Guo, Q., Wu, J., Qiu, X., Huang, X.: Do large language models know what they don't know? arXiv preprint arXiv:2305.18153 (2023)

# QUITO: Accelerating Long-Context Reasoning Through Query-Guided Context Compression

Wenshan Wang[1], Yihang Wang[2], Yixing Fan[1(✉)], Huaming Liao[1],
and Jiafeng Guo[1]

[1] Institute of Computing Technology, Chinese Academy of Sciences, Beijing, China
fanyixing@ict.ac.cn
[2] Beijing University of Posts and Telecommunications, Beijing, China

**Abstract.** In-context learning (ICL) capabilities are foundational to the success of large language models (LLMs). Recently, context compression has attracted growing interest since it can largely reduce reasoning complexities and computation costs of LLMs. In this paper, we introduce a novel Query-gUIded aTtention cOmpression (QUITO) method, which leverages attention of the question over the contexts to filter useless information. Specifically, we take a trigger token to calculate the attention distribution of the context in response to the question. Based on the distribution, we propose three different filtering methods to satisfy the budget constraints of the context length. We evaluate the QUITO using two widely-used datasets, namely, NaturalQuestions and ASQA. Experimental results demonstrate that QUITO significantly outperforms established baselines across various datasets and downstream LLMs, underscoring its effectiveness. Our code is available at https://github.com/Wenshansilvia/attention_compressor.

**Keywords:** Context Compression · In-context Learning · Large Language Model

## 1 Introduction

In recent years, LLMs has demonstrated notable reasoning and generating capabilities, significantly enhancing the performance of natural language processing (NLP) tasks [4]. However, these models still exhibit limitations in acquiring real-time information and integrating external knowledge [8]. In-context learning (ICL) addresses these deficiencies by including examples and relevant contexts directly within the prompts [6]. This approach boost the performance of LLMs in downstream tasks without requiring additional training.

To better improve the reasoning ability of LLMs, researchers propose different ways to incorporate complex contexts in the input [4,8]. For example, retrieval-augmented generation (RAG) employs an additional searcher to retrieve

X. He et al. (Eds.): CCIR 2024, LNCS 15418, pp. 136–148, 2025.
https://doi.org/10.1007/978-981-96-1710-4_11

external relevant documents about the question as the context of inputs, which has attracted lots of attention for both the academia and industry [2,3,8]. In addition, Brown et al. [4] found that the number of examples has a great impact to the reasoning performance of LLMs, where more examples tend to bring better performances [17]. Moreover, the chain-of-thought (CoT) [21,24] further improves the LLMs by involving the reasoning step of each example in the context. While these strategies have the potential to significantly improve the capabilities of LLMs, they also introduce challenges associated with the increased context length, such as higher inference complexity and costs.

To mitigate this issue, context compression in ICL is becoming a prominent solution. On one hand, reducing the length by removing noise from contexts can improve inference efficiency [11,26]. On the other hand, it meets the input length restrictions of open-source LLMs [20,27] while also reduces the costs associated with accessing proprietary LLMs. Several methods [10,11] have been proposed to compress context by estimating the information entropy. This assessment is conducted by utilizing a small external LLM to evaluate the perplexity of individual tokens to identify those that contribute minimal information gain. Tokens that demonstrate low information are subsequently compressed or eliminated. However, neglecting the query during compression may result in the inadvertent deletion of key information.

For the above problem, recent methods such as LongLLMLingua [9] adopt a query-aware compression approach by calculating the perplexity of the context conditioned on the query. Despite this advancement, misalignment between compression model and generation model can lead to inconsistencies in determining which tokens are considered to have "low entropy gain". This discrepancy arises because models may differ in their interpretation and processing of the same information. Our work also scores tokens based on their relevance to the query. However, distinctively, we employ attention metrics rather than perplexity to assess the importance of tokens.

This paper introduces the Query-gUIded aTtention cOmpression (QUITO) method, which strategically selects the context to maintain supporting information by utilizing the attention mechanism. Intuitively, the attention mechanism offers a direct method for analyzing the interactions between the question and the context, moving beyond the sole reliance on models' probabilistic uncertainty. This technique facilitates a more precise identification of the information that is most crucial to the current task. More importantly, the attention-based filtering can be implemented with small LLMs, which improves the computation efficiency.

The main contributions of this study include:

1. This paper proposes a novel context compression method, named QUITO. It utilises self-attention mechanism of Transformers to score the importance of tokens, selecting context relevant to the current query.
2. In contrast to earlier methods that requires a compression model with 7 billion or 13 billion parameters, this method achieves superior results using a smaller LLM with only 0.5 billion parameters.

3. We conduct extensive experiments on two benchmark datasets, which demonstrate the effectiveness of the proposed QUITO. For example, it surpasses strong baselines with an increase in accuracy of up to 20.2.

## 2   Related Work

In this section, we briefly review two lines of related works, i.e., context compression task and attention mechanism.

### 2.1   Context Compression Task

To reduce the length of context, earlier efforts [13] opted to summarize and condense retrieved documents using models such as GPT. Other studies [1, 15, 23, 25] focused on distinguishing between useful and redundant information within documents, training a model to extract the most valuable sentences. For example, LeanContext [1] and FILCO [23] train the model to perform sentence-level extraction for the context. Fit-RAG [15] scores sub-paragraphs with sliding context windows. RECOMP [25] uses a generative model to rewrite extracted candidate sentences, thereby ensuring the coherence and naturalness of the summaries.

Approaches that generate summaries do not allow direct control over the compression ratio, resulting in a growing attention on token and word-level compression techniques in recent times. SelectiveContext [11] utilizes self-information within context for token selection. This approach considers perplexity (PPL) to be the representation of the uncertainty of an LLM regarding information carried by contexts. Based on [11], LLMLingua [10] introduces a two-stage, coarse to a fine, compression method. However, these methods fail to consider the relationship between the context and the query. LLMLingua [9] further addresses this gap by calculating context-specific perplexity conditioned on the query.

The aforementioned token-level compression methods utilize perplexity as the primary filtering criterion. However, discrepancies often arise between smaller compression models and larger generation models in their assessments of word perplexity, making it challenging to align their judgments on lexical importance.

### 2.2   Attention Mechanism

Attention is a significant breakthrough in deep learning, particularly shines in NLP tasks such as translation and summary generation [5]. The core concept behind Attention mechanisms involves assigning a specific weight to each input element, such as words or tokens, indicating their relevance to the task at hand. This allows models to focus selectively on more pertinent parts of the input data.

Self-attention, a particular category of the attention mechanism, measures the relationships between all input elements, assessing how each element influences and relates to the others [16]. Multi-head attention is a key component of the Transformers [22], which improves the model's capability in capturing diverse correlation patterns. Recent studies try to use the attention mechanisms within LLMs to accomplish specific tasks. For instance, DRAGIN [19] use attention to evaluate the extent to which a given text segment significantly influences subsequent content. It employs the perplexity of tokens to determine whether to trigger re-retrieval and regeneration processes. In this paper, we also employ the multi-head attention mechanism to calculate the weights of tokens in context, thereby identifying useless content for answer generation.

## 3    Method

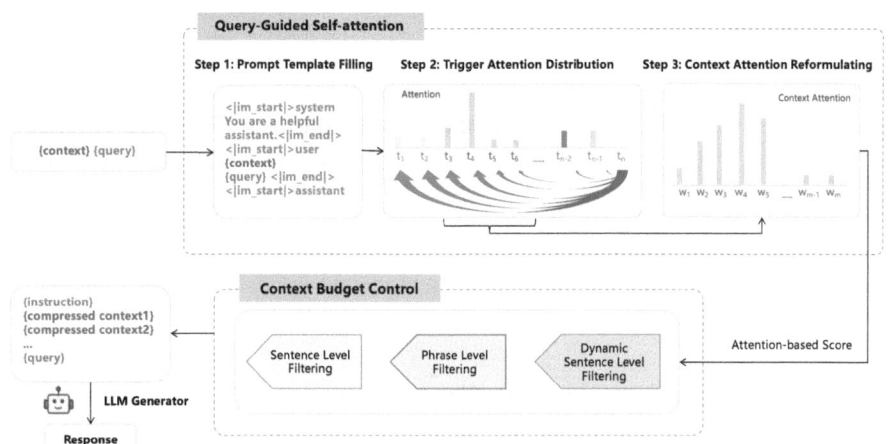

**Fig. 1.** The overall framework of QUITO

In this section, we introduce the QUITO method in detail. As illustrated in Fig. 1, QUITO primarily consists of two main components, namely *the query-guided self-attention* component and *the context budget control* component. In what follows, we will firstly give a formal definition of the task, and then describe each component in detail.

### 3.1    Problem Formulation

Given an input $p = (s, C, q)$, where $s$ is the instruction, $q$ is the query, and $C = \{c_i\}_{i=1}^{n}$ is the context consisting $n$ documents. Every document $c_i = \{w_{i,j}\}_{j=1}^{L_i}$ contains $L_i$ word. The objective of context compression task can be formulated as:

$$\min_{\tilde{C}} dist(P(\tilde{y}|s, \tilde{C}, q), P(y|s, C, q)), \tag{1}$$

where $\tilde{y}$ represents the predicted response of the LLM, and $y$ is the ground truth response. $dist(\cdot, \cdot)$ is a function that measures the distance of two distributions, such as KL divergence. $\tilde{C} = \{\tilde{c}_i\}_{i=1}^{n}$ is the compressed context, and $\tilde{c}_i = \{w_j | w_j \in c_i\}_{j=1}^{\tau L_i}, \tau \in [0, 1]$. $\tilde{c}_i$ is $c_i$ being compressed with ratio at $1/\tau$, where smaller $\tau$ means higher compression ratio.

## 3.2   Query-Guided Self-attention

The query-guided self-attention component aims to estimate the importance of tokens in context by calculating the trigger attention distribution. Firstly, we organize all input by *prompt template filling*. Then, we calculate the importance of all the input with *trigger attention distribution*. Finally, we obtain the lexical units importance within the context by *context attention reformulating*.

**Prompt Template Filling.** It is crucial that the compression model fully understands the task at hand and accurately identifies the information most pertinent to the current query. A standard approach involves concatenating the context with the query and subsequently analyzing how tokens within the query attend to tokens in the context. However, in a Transformer decoder-only architecture, the visibility range of each token in the query varies. This variability suggests that tokens positioned later in the sequence more precisely reflect the model's comprehensive understanding of the task. Given the challenges associated with appropriately weighting tokens at different positions, we propose a novel method that utilize a conversational template and identify a specialized token that encapsulates the compression model's overall understanding of the task.

**Trigger Attention Distribution.** We embed the context and query into a conversational template, concluding with a signal that prompts the model to initiate response generation. The terminal token within this sequence is designated as a trigger token, serving as an indicator of the model's assessment of information need after comprehensively understanding the task at hand. Subsequently, we employ a compression model equipped with a multi-head self-attention mechanism to process the completed template and compute the attention that the trigger token accords to the preceding text:

$$\{\alpha_i | \alpha_i = \frac{\exp(q_{L_{total}}^T k_i)}{\sum_{j=1}^{L_{total}} \exp(q_{L_{total}}^T k_j)}\}, \tag{2}$$

where $q_i$ and $k_i$ are query embedding and key embedding of the *ith* token, respectively. $L_{total}$ is the total number of tokens in the completed template.

**Context Attention Reformulating.** Once the attention allocated by the trigger token to all preceding tokens in the sequence has been determined, the subsequent step involves transforming this attention data into a quantified measure of significance for the lexical units within the context.

The array $\{\alpha_i\}$ signifies attention weights, with its length equating to the aggregate of the lengths of the conversational template, the context, and query. Within the scope of this task, it is imperative to concentrate on the attention distributed to the context segment. The attention should not be diluted by the segments pertaining to the template and the query. Consequently, we implement a normalization process, which is designed to ensure that the distribution of attention across various tokens in the context remains unbiased, robust to the disparities in context and query lengths that may exist across different tasks. For the normalization we use softmax function:

$$\alpha_i' = \frac{\exp(\alpha_{i+doc_{start}})}{\sum_{j=doc_{start}}^{doc_{end}} \exp(\alpha_j)}, i \in [1, doc_{end} - doc_{start}], \tag{3}$$

where $doc_{start}$ and $doc_{end}$ represent the starting and ending positions, respectively, of the context segment.

We consider words to be the smallest semantic units within a document. In order to perform selection on semantic units, the next step involves transforming scores on token to scores attributed to each individual unit. In other words, we need to transform $\{\alpha_i'\}_{i=1}^{L_{doc}}$ to $\{\alpha_i''\}_{i=1}^{L}$, where $L_{doc}$ is the length of token array $\{t_i\}_{i=1}^{L_{doc}}$ that belongs to context, and $L$ is the length of word array $\{w_i\}_{i=1}^{L}$.

A word $w_i$ may consist of one or more tokens. We can formulate a word as $w_i = \{t_j\}_{j=k+1}^{k+l}$, each of which has attention score:

$$\alpha_i'' = \max_{k+1 \leq j \leq k+l} \alpha_j', \tag{4}$$

where the length of the array $\{\alpha_i''\}_{i=1}^{L}$ is $L$.

### 3.3   Context Budget Control

In the previous section, we have derived a list of words, represented as $\{w_i\}_{i=1}^{L}$, and the corresponding array of attention weights, $\{\alpha_i''\}_{i=1}^{L}$. This section introduces the filtering methods that satisfy the requirement of the context budget control.

**Phrase Level Filtering.** In the process of selecting based on attention scores, it is common to inadvertently overlook words adjacent to those with high attention, referred to as target words, which may also contain crucial knowledge for answering the query. To rectify this oversight and ensure these adjacent words are also considered, we apply a weighted adjustment, allowing them to receive a portion of the attention attributed to the target words. This is accomplished by implementing a Gaussian filter across the word attention array $\{\alpha_i''\}_{i=1}^{L}$.

$$G(x) = \frac{1}{2\pi\sigma^2} \exp(-\frac{x^2}{2\sigma^2}) \tag{5}$$

After the application of the Gaussian function $G(x)$ to $\{\alpha_i''\}_{i=1}^{L}$, the resulting Gaussian-modulated attention array is denoted as $\{\alpha_i'''\}_{i=1}^{L}$.

Subsequently, we identify the words from set $\{w_i\}_{i=1}^{L}$ that rank within the top $\tau L$ based on their attention scores $\{\alpha_i'''\}_{i=1}^{L}$.

1. Perform a sort on $\{\alpha_i'''\}_{i=1}^{L}$ on descending order, which yields an ordered set of indices $\{j_1, j_2, \ldots, j_L\}$.

2. Select corresponding words from $\{w_i\}_{i=1}^{L}$ with index $\{j_1, j_2, \ldots, j_{\tau L}\}$, which yields $\{w_i'\}_{i=1}^{\tau L}$.
3. Reorganize the set of selected words $\{w_i'\}_{i=1}^{\tau L}$ to reflect their original sequential order within the context.

Although the selection process targets individual words, the application of Gaussian filtering often leads to the selection of contiguous words, thereby effectively forming phrases.

**Sentence Level Filtering.** In addition to phrase-level filtering, sentence-level filtering is also implemented to preserve more comprehensive semantic information. Using the Natural Language Toolkit (NLTK) toolkit, we extract semantic units at the sentence level. Each sentence $s_i$, denoted as $s_i = \{w_j\}_{j=k+1}^{k+l}$, is assigned an attention score based on the maximum score of the tokens it contains. Subsequently, mirroring the phrase-level filtering process, we prioritize incorporating sentences with higher attention scores into the selection set, while ensuring that the aggregate word count remains below $\tau L$.

**Dynamic Sentence Level Filtering.** Sentence-level filtering often leads to a compression ratio greater than the designated target $1/\tau$. To more effectively adhere to the predetermined compression rate and optimize budget utilization, we augment the results of sentence-level filtering with word-level filtering. Specifically, subsequent to sentence-level filtering, if the count of words is $L'$, we are then able to select an additional $\tau L - L'$ words. These additional words are chosen via phrase-level filtering from the text that was not previously selected. The newly selected words are subsequently concatenated with the results from sentence-level filtering to form the final compressed output.

## 4    Experiments

### 4.1    Datasets and Evaluation Metrics

In this paper, we assess the efficacy of the proposed QUITO method across two distinct scenarios: open domain question answering and long-form question answering. Specifically, we employ the NaturalQuestions (NQ) and ASQA datasets as the testbed.

For NQ dataset, We employed a processed version as described in [14], where each query is paired with 20 documents, among which only one document contains the correct answer. In alignment with the procedures specified in [14], accuracy was used as the metric to determine whether the generated responses accurately included the correct answer. For the ASQA dataset, the answer to the question maybe multi-facet as there are many ambiguous questions. Each ambiguous question in the ASQA dataset has answers reflecting multiple interpretations of these ambiguities. We utilize the dataset version provided by [7], which includes 5 retrieved documents/snippets from Wikipedia for each query. In accordance with [18], our evaluation metrics included Exact Match (EM), a RoBERTa-based QA score (DisambigF1), and ROUGE [12].

**Table 1.** Experimental results of various compression methods applied at different compression ratios on the NaturalQuestions and ASQA datasets.

| Methods | NQ | ASQA | | |
|---|---|---|---|---|
| | Accuracy | RougeL | EM | Disambig_F1 |
| *ratio = 2x* | | | | |
| Selective-Context | 53.2 | - | - | - |
| LLMLingua | 38.7 | 21.3 | 34.6 | 22.2 |
| LongLLMLingua† | 41.2 | 21.6 | 29.7 | 21.2 |
| QUITO (Sentence Level) | 49.9 | 23.5 | **40.3** | 23.6 |
| QUITO (Dynamic Sentence Level) | 58.3 | **23.5** | 40.0 | **23.8** |
| QUITO (Phrase Level) | **58.9** | 21.6 | 38.3 | 22.8 |
| *ratio=4x* | | | | |
| Selective-Context | 38.2 | - | - | - |
| LLMLingua | 32.1 | 20.9 | 33.2 | 21.1 |
| LongLLMLingua† | 33.6 | 20.9 | 24.2 | 20.2 |
| QUITO (Sentence Level) | 52.1 | 22.1 | 30.1 | 20.2 |
| QUITO (Dynamic Sentence Level) | **53.1** | **22.5** | **36.7** | **22.5** |
| QUITO (Phrase Level) | 50.7 | 20.8 | 34.7 | 21.5 |
| Original (without compression) | 68.6 | 23.0 | 45.7 | 26.2 |

## 4.2    Baselines and Implementation

**Baselines.** We take three state-of-the-art compression approaches as baselines: For query-unaware methods, we select Selective-Context [11] and LLMLingua [10], which implements cross entropy scoring to remove redundant vocabulary. For query-aware method, we compare our approach with Longllmlingua [9]. LongLLMLingua implements a two-stage compression method. It first evaluates and reranks multiple retrieved contexts, followed by a token-level compression stage, allocating varying compression budgets to these contexts based on their initial scores. For fair comparison, we excluded the context reranking phrase of LongLLMLingua (marked as LongLLMLingua† in Table 1 and Fig. 2), concentrating on the token-level compression.

**Detailed Implementation.** For fair comparison, we follow LLMLingua [10] to use Longchat-13B-16k[1] as the generation model. To ensure the reproducibility of the results, we apply greedy decoding strategy throughout the inference

---
[1] https://huggingface.co/lmsys/longchat-13b-16k.

process, with the temperature parameter set to zero. The compression model is implemented with Qwen2-0.5B-Instruct[2].

### 4.3   Main Results

Table 1 presents the comparative performance of our method, QUITO, against three baseline methods across various compression rates and datasets. Firstly, we can see that selective-context is a strong baseline compared with LLMLingua and LongLLMLingua† on both $2x$ and $4x$ compression rates. Secondly, QUITO obtains significantly better performances than all baselines, e.g., the improvement of QUITO with phrase level filtering against selective-context, LLMLingua, and LongLLMLingua† (i.e., $2x$ compression ratio) on NQ is 5.7, 20.2, and 17.7, respectively. Finally, we find that QUITO with different filtering method all achieve better performances on both datasets. However, there is no consistent advantages of each filtering method when compared on different datasets. This maybe that the context length on NQ and ASQA differs significantly, i.e., the average length of context on NQ and ASQA is about 2904 and 721 tokens, respectively. All the results demonstrate the effectiveness of QUITO in compressing contexts for the LLMs.

### 4.4   Analysis on Different Position of the Ground Truth Context

We analyse the performance of the QUITO compression method across different ground truth context positions within the NQ dataset. This dataset comprises 20 context document fragments per query, of which only one contains the answer and is designated as the ground truth document. We assessed the impact of this document's positioning at the 1st, 5th, 10th, 15th, and 20th ranks on the efficacy of various compression strategies.

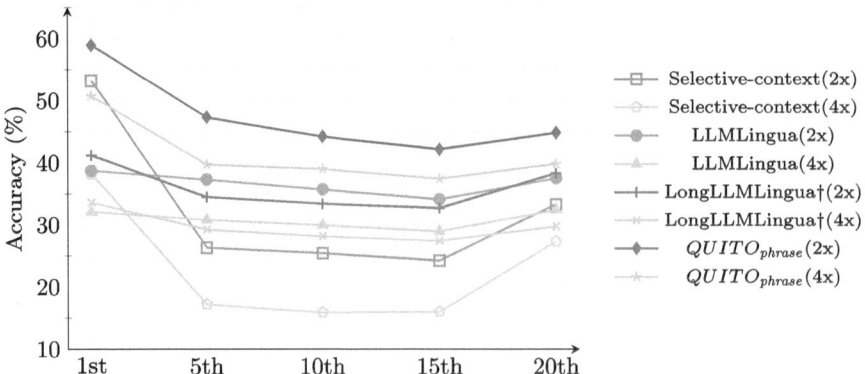

**Fig. 2.** Experimental comparison of different ground-truth context positions.

---

[2] https://huggingface.co/Qwen/Qwen2-0.5B-Instruct.

The results presented in Fig. 2 indicate that all context compression methods struggle with the 'lost in the middle' phenomenon, as described by [14]. Performance is optimal when the ground truth context is positioned at the beginning; however, it deteriorates significantly when the ground truth context is placed in the middle. Among the evaluated methods, LLMLingua [10] exhibits the most resilience to the 'lost in the middle' phenomenon. This robustness may be attributed to its strategy of allocating higher compression ratios to contexts containing a greater density of information. Overall, the QUITO method consistently surpasses the two baseline methods across a variety of ground truth context positions and compression rates. On average, QUITO improves upon the performance of Selective Context [11] by +19.6 and LLMLingua [10] by +13.6.

## 4.5   Analysis on Different Generation Models

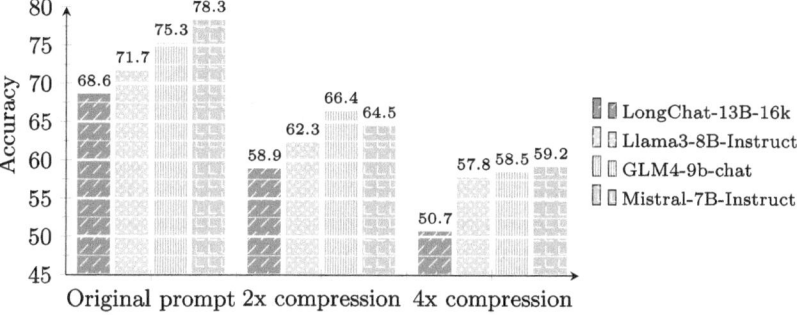

**Fig. 3.** Experimental results of different generation models on NQ dataset.

To better understand the generation ability of different LLMs, we evaluate the performance of 4 widely-used models, including Longchat-13B-16k[3], Llama3-8b-Instruct[4], GLM4-9b-chat[5], and Mistral-7b-instruct[6]. These models were tested with contexts compressed at a rate of 2 on the NQ dataset. The generated responses from these compressed contexts were then compared with those derived from uncompressed contexts.

As depicted in Fig. 3, the Mistral-7B-Instruct model significantly outperforms the other three generation models despite having fewer parameters. This superior performance may be attributed to the incorporation of Grouped-Query Attention (GQA) and Sliding Window Attention (SWA) during its training phase, which enhances its capability to process long sequence inputs. While the context is

---

[3] https://huggingface.co/lmsys/longchat-13b-16k.
[4] https://huggingface.co/meta-llama/Meta-Llama-3-8B-Instruct.
[5] https://huggingface.co/THUDM/glm-4-9b-chat.
[6] https://huggingface.co/mistralai/Mistral-7B-Instruct-v0.2.

compressed at 2x ratio, we find that the GLM4-9b-chat model show the smallest performance decline, with a decrease of 8.9, and the Mistral-7B-Instruct has the greatest decline. When the compression ratio is 4x, we can see that all generation models obtain a relative close performance except for LongChat-13B-16k. These maybe that the LongChat-13B-16k is released earlier than other three models, and the latter are trained more deeply.

## 5   Conclusion

This paper introduces the QUITO method, a novel attention-based importance estimation for long context compression in LLMs. The QUITO method employs a trigger token that comprehensively considers the query to assess the importance of each lexical unit within the context, thereby filtering out units with low relevance scores. Evaluations conducted on the NQ and ASQA datasets demonstrate that our method outperforms state-of-the-art compression methods such as Selective Context, LLMLingua, and LongLLMLingua, confirming its superior ability to preserve essential information needed by LLMs to respond to queries effectively. For future work, we would like to study the combination of the context compression and re-ranking module, since the re-ranking stage in RAG also targets on selecting useful information for final answer generation.

**Acknowledgments.** This work was funded by the National Natural Science Foundation of China (NSFC) under Grants No. 62372431, the Strategic Priority Research Program of the CAS under Grants No. XDB0680102, the National Key Research and Development Program of China under Grants No. 2023YFA1011602, the Youth Innovation Promotion Association CAS under Grants No. 2021100, the Lenovo-CAS Joint Lab Youth Scientist Project, and the project under Grants No. JCKY2022130C039.

**Disclosure of Interests.** The authors have no competing interests to declare that are relevant to the content of this article.

## References

1. Arefeen, M.A., Debnath, B., Chakradhar, S.: Leancontext: cost-efficient domain-specific question answering using llms (2023). https://arxiv.org/abs/2309.00841
2. Asai, A., Wu, Z., Wang, Y., Sil, A., Hajishirzi, H.: Self-rag: learning to retrieve, generate, and critique through self-reflection (2023). https://arxiv.org/abs/2310.11511
3. Borgeaud, S., et al.: Improving language models by retrieving from trillions of tokens (2022). https://arxiv.org/abs/2112.04426
4. Brown, T.B., et al.: Language models are few-shot learners. arXiv preprint arXiv:2005.14165 (2020)
5. Chaudhari, S., Mithal, V., Polatkan, G., Ramanath, R.: An attentive survey of attention models (2021). https://arxiv.org/abs/1904.02874
6. Dong, Q., et al.: A survey on in-context learning (2024). https://arxiv.org/abs/2301.00234

7. Gao, T., Yen, H., Yu, J., Chen, D.: Enabling large language models to generate text with citations. In: Bouamor, H., Pino, J., Bali, K. (eds.) Proceedings of the 2023 Conference on Empirical Methods in Natural Language Processing, pp. 6465–6488. Association for Computational Linguistics, Singapore, December 2023. https://doi.org/10.18653/v1/2023.emnlp-main.398. https://aclanthology.org/2023.emnlp-main.398

8. Gao, Y., et al.: Retrieval-augmented generation for large language models: a survey (2024)

9. Jiang, H., et al.: Longllmlingua: accelerating and enhancing llms in long context scenarios via prompt compression. ArXiv preprint abs/2310.06839 (2023). https://arxiv.org/abs/2310.06839

10. Jiang, H., Wu, Q., Lin, C.Y., Yang, Y., Qiu, L.: LLMLingua: compressing prompts for accelerated inference of large language models. In: Bouamor, H., Pino, J., Bali, K. (eds.) Proceedings of the 2023 Conference on Empirical Methods in Natural Language Processing, pp. 13358–13376. Association for Computational Linguistics, Singapore, December 2023. https://doi.org/10.18653/v1/2023.emnlp-main.825. https://aclanthology.org/2023.emnlp-main.825

11. Li, Y., Dong, B., Lin, C., Guerin, F.: Compressing context to enhance inference efficiency of large language models (2023)

12. Lin, C.Y.: ROUGE: a package for automatic evaluation of summaries. In: Text Summarization Branches Out, pp. 74–81. Association for Computational Linguistics, Barcelona, Spain, July 2004. https://aclanthology.org/W04-1013

13. Liu, J., Jin, J., Wang, Z., Cheng, J., Dou, Z., Wen, J.R.: Reta-llm: a retrieval-augmented large language model toolkit (2023). https://arxiv.org/abs/2306.05212

14. Liu, N.F., et al.: Lost in the middle: How language models use long contexts (2023). arXiv:2307.03172

15. Mao, Y., Dong, X., Xu, W., Gao, Y., Wei, B., Zhang, Y.: Fit-rag: black-box rag with factual information and token reduction. ACM Trans. Inf. Syst., July 2024. https://doi.org/10.1145/3676957. Just Accepted

16. de Santana Correia, A., Colombini, E.L.: Attention, please! a survey of neural attention models in deep learning (2021). https://arxiv.org/abs/2103.16775

17. Song, Y., Wang, T., Mondal, S.K., Sahoo, J.P.: A comprehensive survey of few-shot learning: evolution, applications, challenges, and opportunities (2022). https://arxiv.org/abs/2205.06743

18. Stelmakh, I., Luan, Y., Dhingra, B., Chang, M.W.: ASQA: factoid questions meet long-form answers. In: Goldberg, Y., Kozareva, Z., Zhang, Y. (eds.) Proceedings of the 2022 Conference on Empirical Methods in Natural Language Processing, pp. 8273–8288. Association for Computational Linguistics, Abu Dhabi, United Arab Emirates, December 2022. https://doi.org/10.18653/v1/2022.emnlp-main.566. https://aclanthology.org/2022.emnlp-main.566

19. Su, W., Tang, Y., Ai, Q., Wu, Z., Liu, Y.: Dragin: dynamic retrieval augmented generation based on the real-time information needs of large language models. ArXiv abs/2403.10081 (2024). https://api.semanticscholar.org/CorpusID:268509926

20. Touvron, H., et al.: Llama: open and efficient foundation language models (2023). https://arxiv.org/abs/2302.13971

21. Trivedi, H., Balasubramanian, N., Khot, T., Sabharwal, A.: Interleaving retrieval with chain-of-thought reasoning for knowledge-intensive multi-step questions (2023). https://arxiv.org/abs/2212.10509

22. Vaswani, A., Shazeer, N., Parmar, N., Uszkoreit, J., Jones, L., Gomez, A.N., Kaiser, L., Polosukhin, I.: Attention is all you need. In: Proceedings of the 31st Interna-

tional Conference on Neural Information Processing Systems (NeurIPS), pp. 6000–6010 (2017)

23. Wang, Z., Araki, J., Jiang, Z., Parvez, M.R., Neubig, G.: Learning to filter context for retrieval-augmented generation (2023). https://arxiv.org/abs/2311.08377

24. Wei, J., et al.: Chain-of-thought prompting elicits reasoning in large language models (2023). https://arxiv.org/abs/2201.11903

25. Xu, F., Shi, W., Choi, E.: Recomp: Improving retrieval-augmented lms with compression and selective augmentation. ArXiv abs/2310.04408 (2023). https://api.semanticscholar.org/CorpusID:263830734

26. Xu, Z., et al.: Compress, then prompt: improving accuracy-efficiency trade-off of llm inference with transferable prompt (2023). https://arxiv.org/abs/2305.11186

27. Yang, A., et al.: Baichuan 2: open large-scale language models (2023). https://arxiv.org/abs/2309.10305

# Author Index

X. He et al. (Eds.): CCIR 2024, LNCS 15418, p. 149, 2025.
https://doi.org/10.1007/978-981-96-1710-4